Urban Animals

The city includes opportunities as well as constraints for humans and other animals alike. Urban animals are often subjected to complaints; they transgress geographical, legal, and cultural ordering systems, while roaming the city in what are often perceived as uncontrolled ways. But they are also objects of care, conservation practices, and bio-political interventions. What, then, are the "more-than-human" experiences of living in a city? What does it mean to consider spatial formations and urban politics from the perspective of human/animal relations?

This book draws on a number of case studies to explore urban controversies around human/animal relations, in particular companion animals: free-ranging dogs, homeless and feral cats, urban animal hoarding, and "crazy cat ladies." The book explores "zoocities," the theoretical framework in which animal studies meets urban studies, resulting in a reframing of urban relations and space. Through the expansion of urban theories beyond the human, and the resuscitation of sociological theories through animal studies literature, the book seeks to uncover the phenomenon of "humanimal crowding," both as threats to be policed and as potentially subversive. In this book, a number of urban controversies and crowding technologies are analyzed, finally pointing at alternative modes of trans-species urban politics through the promises of humanimal crowding—of proximity and collective agency. The exclusion of animals may be an urban ideology, aiming at social order, but close attention to the level of practice reveals a much more diverse, disordered, and perhaps disturbing experience.

Tora Holmberg is an Associate Professor in Sociology and Senior Lecturer at the Department of Sociology and the Institute for Housing and Urban Research, Uppsala University, Sweden.

Routledge Human–Animal Studies Series
Series Edited by Henry Buller, Professor of Geography,
University of Exeter, UK

The new *Routledge Human–Animal Studies Series* offers a much-needed forum for original, innovative and cutting-edge research and analysis to explore human–animal relations across the social sciences and humanities. Titles within the series are empirically and/or theoretically informed, and explore a range of dynamic, captivating and highly relevant topics, drawing across the humanities and social sciences in an avowedly interdisciplinary perspective. This series will encourage new theoretical perspectives and highlight groundbreaking research that reflects the dynamism and vibrancy of current animal studies. The series is aimed at upper-level undergraduates, researchers and research students as well as academics and policy-makers across a wide range of social science and humanities disciplines.

Published

Critical Animal Geographies
Politics, intersections and hierarchies in a multispecies world
Edited by Kathryn Gillespie and Rosemary-Claire Collard

Urban Animals
Crowding in zoocities
Tora Holmberg

Forthcoming

Animal Housing and Human–Animal Relations
Politics, practices and infrastructures
Edited by Kristian Bjørkdahl and Tone Druglitrø

Taxidermy and Contemporary Art
Giovanni Aloi

Urban Animals

Crowding in zoocities

Tora Holmberg

Routledge
Taylor & Francis Group

LONDON AND NEW YORK

First published in paperback 2017

First published 2015
by Routledge
2 Park Square, Milton Park, Abingdon, Oxon OX14 4RN

and by Routledge
711 Third Avenue, New York, NY 10017

Routledge is an imprint of the Taylor & Francis Group, an informa business

British Library Cataloguing in Publication Data
A catalogue record for this book is available from the British Library

Library of Congress Cataloging in Publication Data
Holmberg, Tora.
Urban animals : crowding in zoocities / by Tora Holmberg.
pages cm. — (Routledge human-animal studies series)
Includes bibliographical references.
ISBN 978-1-138-83288-6 (hardback) — ISBN 978-1-315-73572-6
    (e-book) 1. Urban animals. 2. Human-animal relationships. I. Title.
QH541.5.C6H65 2015
591.75'6—dc23
2014038199

ISBN: 978-1-138-83288-6 (hbk)
ISBN: 978-1-138-63505-0 (pbk)
ISBN: 978-1-315-73572-6 (ebk)

Typeset in Times New Roman
by Swales & Willis, Exeter, Devon, UK

To Ronja and Rocky, my guides to zoocities encounters

Contents

List of figures   x
Notes on the contributors   xi
Acknowledgments   xii

1 Urban animals   1

PART I
Animals in the city   **21**

2 Bodies on the beach: Allowability and the politics of place 23
3 Stranger cats: Homelessness and ferality in the city   47

PART II
Humanimal transgressions   **69**

4 Verminizing: Making sense of urban animal hoarding   71
5 Feline femininity: Emplacing cat ladies   97

PART III
The promises of crowding in zoocities   **117**

6 Beyond crowd control   119
7 Open endings   128
   CO-AUTHORED WITH KATJA AGLERT

Bibliography   147
Index   159

Figures

Frontispiece Zoocities, collage (Katja Aglert 2014) i

1.1	Urban deer, Uppsala	2
1.2	Berlin Zoo, Germany	4
1.3	Lijnbaansgracht, Amsterdam	9
1.4	Urban dogs, Bilbao, Spain	11
2.1	Prohibition sign, Santa Cruz	24
2.2	Dog resting, Berlin	25
2.3	Dog resting, Thailand	26
2.4	Anti-homeless spikes in London	31
2.5	Dogs on beach, Barry Island, UK	33
2.6	Dog on beach, Barry Island, UK	40
3.1	Street cat, Lanzarote	49
3.2	Street cat in China town, Manila, Philippines	67
4.1	"Stray dogs," Kate Ward, Camberley, UK	72
4.2	"Svankvinnan slår till igen" ("The Swan Lady strikes again")	73
4.3	Cat in bed	85
5.1	Cat woman and Batman	98
5.2	Vällingby 1960/ABC City 1952	108
7.1	Balcony on 12th floor, Stockholm	139

Contributors

Tora Holmberg is an Associate Professor and Senior Lecturer at the Department of Sociology and the Institute for Housing and Urban Research, Uppsala University, Sweden. Her expertise lies in the intersection of animal studies, science and technology studies, feminist theory, cultural sociology, and urban studies. She has published books including *Science on the Line* (in Swedish, 2005), *Investigating Human/Animal Relations* (ed., 2009), *Dilemmas with Transgenic Animals* (in Swedish, with Malin Ideland, 2010), and has been published in a wide range of journals, e.g. *Space & Culture*, *Feminist Theory*, *Bio-Societies* and *Public Understanding of Science*, on topics such as animal experimentation, feminist epistemologies, urban politics and human/animal relations. She is also a co-editor of *Humanimalia*.

Katja Aglert is an artist based in Stockholm, Sweden. Her practice is based in interdisciplinary research, and includes both individual and collaborative projects. Her work has been exhibited at Marabouparken (Sweden), the National Museum of Denmark, the 58th International Short Film Festival Oberhausen (Germany), and elsewhere. She is a senior lecturer at the Department of Fine Art at Konstfack University College of Arts, Crafts and Design, Sweden. Publications including her work include *Shaped by Time* (2012) and *Winter Event—antifreeze, Winter Event—antifreeze, Winter Event—antifreeze, Winter Event—antifreeze* (2014).

Acknowledgments

The work with *Urban Animals*, which has now come to an end, goes back several years and has been made possible through the generous contributions of numerous people and institutions. While it is difficult to recall and rehearse everyone, I still want to try, knowing that I will no doubt fail. Please, bear with me . . .

This book would not have been the same without the generous and trusting contributions of interviewees and other informants in the field of urban animal welfare. Moreover, people from near and far have shared their histories of animal hoarding: their own, their neighbors, or relatives. Many thanks for sharing these invaluable narratives, for providing me with insights to human/animal worlds that few of us know much about. I also want to thank the artists who have generously contributed their photos for the book: Christer Barregren, Joao Bento, Michael Heath, and Andrew Horton.

Moreover, I am grateful to the Institute for Housing and Urban Research, Uppsala University, funder of the assistant professor position that enabled the project *Controversial connections. Human/animal relations in the city* (2010–2013). This flourishing environment, including the interdisciplinary research seminar, has been immensely helpful to the success of the task. In particular, I would like to thank Brett Christophers, Helen Ekstam, Mats Franzén, Terry Hartig, Christina Kjerman-Meyer, Kerstin Larsson, Irene Molina, Göran Rydén, Eva Sandstedt, and Cecilia Öst, for providing constructive comments on poor paper drafts, and for general encouragement and collegial support.

Other environments at Uppsala University that I have had the privilege to be part of for longer periods are, first, the research cluster the Humanimal Group at the Centre for Gender Research, Uppsala University, of which I have been a dedicated member since its beginning in 2007. I have met such talented and tremendously sharp scholars within this interdisciplinary environment. In particular, I am dedicated to Jacob Bull, Rebekah Fox, Eva Hayward, Ann-Sofie Lönngren, David Redmalm, Anna Samuelsson, Pär Segerdahl, and Harlan Weaver. You are the best!

Interdisciplinary research aside, my disciplinary home of sociology, and the sociological community of Uppsala University and beyond, have been immensely important in sharpening the theoretical questions at stake and in encouraging me to proceed in experimental and difficult directions. The Department of Sociology,

Uppsala University, where I am now again a member of the faculty, has been an important anchor, not least since most of my PhD students reside there. In particular, I want to thank Kalle Bergren, Maria Eriksson, Hedda Ekerwald, Erik Hannerz, Agneta Hugemark, Clara Iversen, Fredrik Palm, Keith Pringle, Lennart Räterlinck, and many more. The Cultural Sociology network in Sweden, led by Anna Lund and Mats Trondman, is also one such highly competent and generous group of scholars of which I have had the privilege to be a part. Thank you, all of you! Other Swedish sociologists, such as Hanna Bertilsdotter, Kerstin Jacobsson, Simon Lindgren, and Maria Törnqvist, have been important to this process, by commenting on the work at various stages of progress (or lack thereof).

I have benefited from visiting many research groups and environments internationally during the time of the research project. Donna Haraway and the History of Consciousness Department, University of California, Santa Cruz, generously invited me for a visiting research position in 2010. Rowland Atkinson and Simon Parker arranged a fruitful seminar at CURB, York University, UK, in 2011. I am also indebted to Sophie Watson and others within the Urban Ecologies network at the Open University, Milton Keynes, UK, for sharing their knowledge for a few days in 2013. Individual scholars within the global network of animal studies, who have been invaluable for the progression of this project are, among others not mentioned, Kristian Björkdal, Lynda Birke, Tone Druglitrø, Lene Koch, Michael Lundblad, Susan McHugh, Mara Miele, Liv Emma Thorsen, Cecilia Åsberg, and the excellent, brilliant and inspiring *Humanimalia* collective of editors. Many, many thanks to all of you!

Thanks to the publishers, especially the series editor Henry Buller and editor Faye Leerink, for believing in the project. My warm thanks also to the anonymous reviewers.

Last but certainly not least, I am grateful to Katja Aglert, a brilliant artist and intelligent scholar, who has been a truly inspirational support through the work with finalizing the book, in discussing weaknesses and strengths, and pointing out alternative interpretations. Our joint work for Chapter 7 was such a pleasure, trying out an alternative form of collective authoring. Moreover, Katja has provided invaluable contributions to the book in the shape of the figures—as an image editor she has helped to strengthen the analytical points I wanted to make. Interdisciplinary work at its best!

Uppsala, September 5, 2014

***

The author wishes to acknowledge that extracts from the following journal articles and book chapters appear in this book, and to thank the publishers for their kind permission to reproduce this material:

Chapter 2: Holmberg, T. (2013) "Trans-species urban politics. Stories from a beach," *Space and Culture*, 16(1): 28–42.

Chapter 3: Holmberg, T. (2014) "Wherever I lay my cat? Post-human urban crowding and the meaning of home," in Marvin, G., McHugh, S. (eds.) *Routledge Handbook of Human–Animal Studies*, Routledge: New York: 54–67.

Chapter 4: Holmberg, T. (2014b) "Sensuous governance. Assessing urban animal hoarding," *Housing, Theory & Society*, 31(4): 464–479.

Chapter 5: Holmberg, T. (2014) "Kattkvinnor," in Thorslund, C.A., Gjerløff, A.K. (eds.) *Dyrisk og dyrebar—kæledyr i et samfundsvidenskabeligt lys*, Nyt fra Samfundsvidenskaberne: Copenhagen: 129–151.

These articles and chapters have been developed and reframed within the context of the book as a whole.

1 Urban animals

One afternoon in the middle of May, when taking a regular walk with my dogs Ronja and Rocky in the suburban area where we live, a deer suddenly turned up on the path from the trees surrounding her. She looked at us from a distance of a few meters for what seemed like for ever, but in reality was probably no more than a second, and then in haste continued in the direction of a nearby housing area. Needless to say, the dogs—after an initial freeze—got very excited, barking and pulling at their leashes, frustrated and anxious to follow the deer. First, I marveled, caught by the unexpected multi-species meeting and the beauty of the deer. Then, I became anxious too, and held on tightly to the dogs' leashes, fearing that they would manage to break free. Then I got irritated: what was she doing here, disturbing our precious routine? Then relieved: happy that my canine followers had not at that moment been off the leash (in which case, I am certain they would have chased the deer). These emotions followed quickly and seamlessly in a few seconds, framed and created by the interactional, spatial and temporal context. Had we met the deer in the woods outside of the city, or even early in the morning (urban deer typically belong to the night), the meeting would have been experienced differently. It also would have been regulated differently: in Sweden—the national context of the incident—dogs are not allowed to be off-leash (or at least not to run free) in the woods during the period between March 1 and August 20 (Hunting Act 1987: 259; Dog and Cat Supervision Act 2007: 1150). Even at other times and in other places, dogs must always be under voice control if ranging free. But, obviously, neither deer nor dogs care much about city limits (see Figure 1.1).

This brief moment illustrates the main question targeted in this book: namely, what are the multi-species experiences and politics of living in a city? Cities are inhabited by an array of species, all contributing to urban life and its spirit (Wolch 2002). The world is also getting more and more urbanized, with a growing proportion of people and other animals living in cities. The reasons for multi-species urbanization, while complex, "include two major constellations of causes: animals [as well as humans] are choosing to move into city spaces, and animals are finding their homes overtaken by cities" (van Dooren and Rose 2012: 1). So-called wild animals—mammals like foxes, rats,

Figure 1.1 Urban deer, Uppsala (Photo: Tora Holmberg 2013)

and the deer above—populate cities, as do many kinds of birds, fish, invertebrates, reptiles and insects, making up a complex urban ecology (Dion and Rockman 1996). Companion animals follow human migration, and urban cats and dogs, as well as other species, are growing in number. Some live indoors with humans and stay in different versions of human homes, while others occasionally or permanently occupy public spaces, and live in parks and streets. These animals—in particular dogs and cats—and their relationships with humans, are investigated in this book. Companion animals are however not only urbanized, but also more or less domesticated. Domestication is a historical and spatial relational process in which some animals come to co-evolve with humans, for better or for worse (Haraway 2003: 30–31). Urbanized animals, including humans, depend on the city for their livelihood; they suffer and thrive because of their urban lifestyle. But urbanization comes along with conflicts over space: who is allowed where, and under what conditions? Who is involved in processes of "politics of place" and who gets the right to define the meaning and function of a particular place (Franzén 2002)? By "politics," I address the more-than-discursive acts involved in struggles over priority of access, interpretation and resources. In this connection, one further complication is that urban animals are difficult to discipline; they often transgress legal as well as cultural ordering systems, while roaming the city in what appear uncontrolled ways. But they are similarly turned into objects of care, conservation practices, and bio-political interventions. Thus, controversies over animal presences are to be expected when these different "frames of meaning" (Goffman 1974) implode.

Drawing from case studies in a cultural comparative and intersectional approach, this book explores a number of controversies around urban human/animal relations, in particular cat/dog/human ones. The subtitle—Crowding in zoocities—alludes to the theoretical framework in which the multidisciplinary field of animal studies meets urban sociology, resulting in a reframing and a reinterpretation of urban relations and space. The "zoo" of the title refers to combinations of kinds and species of living animals. The term was chosen purposely, in order to emphasize the problematic division between humans and other animals—we are of course all animals and thus also in that sense zoo—and to highlight real animals, animal lives, rather than animals as metaphors or symbols. Moreover, the term "zoo" connotes "a situation characterized by confusion and disorder" (*Oxford English Dictionary* 2012), something that is often true of human/animal encounters. However, the zoo as a particular place also forms an ordered space for such encounters. Like other forms of exhibits of animals, e.g. the diorama of the natural history museum, the zoo tells us stories of particular histories and ideologies, reconstructing nature in particular ways. In Kay Andersson's study of Adelaide Zoo, he demonstrates how, over time, it has shifted meanings from colonial power, to leisure, to education (1997). Today the zoo predominantly speaks of conservation of endangered species and environmental concerns. Common to these narratives, however, is that they have continually reinstated boundaries between humans and other animals, ordering the perceived disordered-ness of nature. Thus "zoo," as a term, refers both to disorder and to an urban place that is set to organize and display it, neutralizing and normalizing the inherent wildness of the zoo animals, so that they become "in place" within human spaces (Philo and Wilbert 2000: 22). Zoo thus embodies contrasts that connote threats and categorization, knowledge and power (see Figure 2.2).

Through the expansion of sociological theories beyond the human perspective, and the resuscitation of sociological theories through animal studies literature, I seek to uncover the phenomenon of human/animal crowding in zoocities, both as a threat to be policed and as a potentially subversive element. The "crowding" in the subheading of this book's title aims at capturing notions of the uncontrolled and spatial transformation of individuals across species, within specific contexts, thus the concept of crowding is in conversation with the notion of zoo. Throughout the book, a number of crowding phenomena and control technologies are analyzed, finally pointing at alternative modes of "multi-species urban politics." To this end, I develop a framework for thinking about urban human/animal relations that involves the dimensions of species and spaces. In addition, I introduce the notion of senses. This is justified as the bridging between species and spaces—we experience certain spatial arrangements sensuously and spaces are formed by bodies.

Empirically, the book examines the politics of place and bodies taking shape in public spaces, as well as the boundaries of what may be considered proper living, e.g. the norms of urban living in late modern consumer society.

Figure 1.2 Berlin Zoo, Germany (Photo: Katja Aglert 2014)

This is done through the study of a number of controversial cases, including unleashed dogs, homeless and feral cats, animal hoarding, and crazy cat ladies. Studies of human/cat/dog relations, if approaching place at all, most often do so with regard to homes or shelters and not urban space, while studies of *urban* animals seldom focus on domestic animals but "wild" ones. What I do in this book is to go against this separation and place domestic animals and humans in the context of the city. Through the studies, the notions of "wild" and "domestic," "pet" and "vermin," "human" and "animal" are set in motion and transformed. As a theoretical undertaking, the book contributes to "more-than-human" urban theory in two ways. First, urban "sociation," where, typically, subjectivity processes, including social interaction and group formation, are thought to take shape in the social realm of humans. What does it mean to become one in a multi-species context of many (Haraway 2008a)? Second, urban space as a process constructed through the dialectics of form and experience. Urban theorists have more or less neglected the presence of animals in the city, and have thought about processes like crowding and urbanization as purely human (Wolch et al. 1995). Challenging this notion, what does it mean to consider spatial formations and urban politics from the perspective of human/animal relations?

The approaches to answering the theoretical questions are manifold. One important background of my study is research in the interdisciplinary field of animal studies, and the work of Donna Haraway is pivotal here. An important point is that becoming human is a multi-species process of co-evolution in all the dimensions of this concept (2003, 2008a). By this she means that human-ness as we know it is an outcome of a historical process of living together— in competition as well as in symbiosis—with other animals. Complementing this approach, I use the work in so-called hybrid geographies to understand urban space and subjectivity as something "more-than-human" (Hinchliffe and Whatmore 2006; Whatmore 2006). The hybrid approach refuses to oppose nature and culture, social and material, object and subject, and instead develops an understanding of the linkages and entanglements of these presupposed bina-ries (Whatmore 2002; Hinchliffe 2007). In the urban context, the construction of the civilized and ordered city rests on the continuous exclusion of nature (Kaika 2005). Urban theorists such as Matthew Gandy (2003), for example, stress that nature has always been an object of urbanization, and show how raw material becomes transformed into "metropolitan nature" by, for example, new water technologies. However, as already stated, I am not interested in urban nature or life—"bios"—in a general or abstract sense, but in human/animal rela-tions in particular. Consequently, I use the concept "more-than-human," which aims at de-stabilizing anthropocentrism and de-centering the category of the human itself (Hinchliffe and Whatmore 2006). The approach offers theoretical guidance, as well as an epistemological and ethical stance: responsible social scientists have an obligation to investigate the world as constituted by human and non-human actors in order to challenge the ideology of human exceptional-ism (Hinchliffe and Whatmore 2006: 136). To this end, I focus on urban politics and bodies. Gandy writes:

> Yet if we are to make sense of the modern city—and its post-industrial, late-modern and post-modern permutations—we need to engage with the body both as a site of corporeal interaction with the physical spaces of the city and as a symbolic field within which different aspects to the legitimation of modern societies are played out.
>
> (Gandy 2006a: 497)

That bodies are simultaneously material and symbolic can be illustrated by the deer episode described above. There and then, human, dog and deer bodies inter-acted and intervened in a physical environment. Moreover, although our bodies and movements are all regulated through human laws and norms, these regula-tions look very different and their effects differ profoundly depending on species. For example, while my human body is legally protected from lethal violence, I am allowed to have my own dogs killed at any time. The deer on the other hand is pro-tected from hunting during most of the year. But even when hunting is allowed, only registered hunters are permitted to hunt, and only in restricted areas—not, for example, in one's garden. Obviously, bodies and matters of power and control go together.

The hybrid, "more-than-human" approach is excellent in showing how modern dichotomies are arbitrarily constructed. However, I am convinced that these new approaches have overlooked some essential insights from urban sociology, related to the dialectics and interconnectedness of subject and form, of interaction and context, of action and politics. Dialectics here means mutually informing, co-shaping through the tensions of concrete, everyday practice (Shields 1998). The dialectical perspective should not, however, be confused with a dualist one; to be sure, hybrid approaches have taught us that dualisms are constructed, they are arbitrary, and work to conserve existing social orders (Haraway 1991). However, the collapsing of categories into new, imploded ones is not analytically satisfactory either. Thus, if one wants to understand actors' experiences of certain places, the interdependence of subjectivity and form gets lost if one focuses solely on hybridity.

The challenge undertaken in this book is to read these more-than-human approaches together with classical sociological traditions as well as with modern urban sociology. As already stated, I focus particularly on the relationship between body and city through the concept of crowding (more on this below), and investigate it through the joint reading of sociological theory and hybridity perspectives. Therefore my approach is to move from animal studies to urban sociology and back again. It is my conviction that animals potentially disrupt disciplinary systems, and that looking through the lens of human/animal relations can transform and improve disciplinary thinking (Wolfe 2009). Animals are in this sense "undisciplined," and the encounters with other animals force us to creatively tweak and expand our ingrained conceptual frames (Segerdahl 2011).

The remainder of this introduction will deal with developing and clarifying the theoretical position in relation to species, spaces and senses. I will lay out the foundation of the concept of "humanimal crowding" and, after this exercise, introduce the concept of "zoo-ethnographies"—the methodologies used—and, finally, outline the chapters of the book.

Species

Urban theory in general is firmly anthropocentric, and the fact that humans are placed at the center of analyses is seldom reflected upon (Wolch et al. 1995). Urbanization has been theorized, as cultural geographer Jennifer Wolch points out, as a process acting on "empty spaces," not taking non-human inhabitants seriously (1998: 119). Typically, studies of urban human/animal relations come from geographers (Wolch and Emel 1998; Philo and Wilbert 2000; Wolch 2002; Hinchliffe and Bingham 2008), historians (Atkins 2012), literary scholars (Mason 2005), or anthropologists (Sabloff 2001). Nevertheless, from

the 1990s onward, there has been an increasing sociological interest in human/
animal relations (Arluke and Sanders 1996; Franklin 1999; Sanders 1999;
Franklin and White 2001; Holmberg 2006; Peggs 2012). Human relations to
other animals are part of the construction of urban identities and places, and
society rests on the use of other animals (Griffiths et al. 2000). Humans and
other animals co-construct emplaced routines, relations, and a shared history
(Jerolmack 2013: 35). One reason for sociologists' reluctance to deal with
human/animal relations is the nature/culture distinction, inherited from classical sociological theory (Myers 2003). While "society" consists of humans and
their culture, animals are placed in "nature" (Sabloff 2001). In the ideology of
modern city life, this division is reinforced since "nature" has no significant
place in the city, perceived as civilized. It is well known that the presence
of other animals can "de-civilize" urban places, which in turn calls for normalizing strategies such as sanitization (Griffiths et al. 2000). But, as Colin
Jerolmack puts it, "the edges of the city and nature continually rub against,
and run over, each other like tectonic plates" (2013: 16). In everyday life,
the boundaries are constantly blurred and reinstated, something that will be
shown throughout the book.

Since cats and dogs—the animals included in this study—occupy what can be
called a "liminal space" of being domestic but not human, family but still animal
(Fox 2006), processes of exclusion and inclusion, when they occur, can be rather
complex. Liminal creatures and other anomalies, as we have learned from Mary
Douglas, can be perceived as threatening cultural order and sorting systems, and
thus trigger closure and other normalizing strategies. However, also through their
multiple meanings, they can work as potentially subversive actors and contribute
to social change (Douglas 1997 [1966]). David Redmalm writes about the notion
of the pet, that it:

> is created at the intersection of categorical borders humans have drawn,
> and it is an accomplice in the reproduction of these borders. However,
> an anomaly is potentially disruptive—it draws attention to and makes us
> aware of the borders along which it moves. Anomalous phenomena are
> not only loved for their unique status—they also threaten an established
> social order. When living beings take the role of anomalous phenomena,
> they are regularly as a consequence exploited, objectified, abused, killed,
> and ridiculed.
>
> (Redmalm 2013: 16–17)

One way of dealing with liminality and boundary-crossing in human/animal
relations is through the lens of post-humanism, which refers to a temporal
space—after humanism—but, more important, to a theoretical space (Wolfe
2009). It asks what happens if we move humans from the center of inquiry. In
short, post-humanism constitutes a challenge to Humanism with a capital H,
which is construed as anthropocentric, and connected to the dominance and

exploitation of nature and other animals. Within Humanism, "human" is a category that you need to qualify for, and thus there are numerous examples of categories that, historically, have fallen outside the category (slaves, women, Jews, refugees, aboriginal people, etc.). Giorgio Agamben (2004) refers to the "anthropological machine," through which the "human" constantly delineates the "beast," and mirrors it in the other in order to define itself. Through similar processes, humans can become animalized, and thus objects for "hygienic" interventions (Agamben 2004). For example, the intellectuals of eighteenth-century Paris saw in public butchering a threat to civilization, in part because of the blurring of species boundaries: the butcher became a beast and the bull a victim (Krefting 2009). Thus, removing animal killing from the streets to slaughterhouses outside the city was the logical solution in the early modern European metropolis (see also Atkins 2012). People interacting too intensely with birds in public, become "birdies" or "pigeon ladies," the animality of the birds spilling over to the human actors. Feral cat feeders may be viewed as discrepant as they invest too much in non-human relations. Another example of particular interest for this book is that people who live in the company of many companion animals seem to challenge certain norms concerning what it is to be human, and risk becoming animalized in popular discourse (see Chapters 4 and 5, see Figure 1.3).

One could say that urban animals work against modernity's purification process, since their ambiguous identities risk polluting clean categories by hybridizing, for example, the dichotomy of nature/culture, wild/domestic, public/private (see also Latour 1993). Haraway's (2003) conceptual development based on the working relation between human and dog, a relation defined by cooperation and responsibility, but in reality often abuse and neglect, offers a way to understand this ambiguity. Pivotal is to understand species as kin, in itself as relational, as Eva Hayward writes:

> Species exist in taxonomic differences (Homo sapiens are not the same as Octopus vulgaris), but species are also always already constitutive of each other through the spaces and places we cohabit—this of course includes language and other semiotic registers. Indeed, species are relationships between species—relationality is world-hood.
>
> (Hayward 2008: 254)

Moreover, species are not related in abstract spaces, they "take place"; humans and other animals encounter and constitute one another in and through particular "contact zones" of mutual, corporeal entanglements (Haraway 2008a: 4). Even though we are not dealing with equal relations, the embracing of what Haraway calls "significant otherness" (2003) may lead to new insights and practices. However, significant otherness also includes its ever present shadow: disconnection. Significant otherness is thus about alliances, conflicts, and distance between species, about being "messmates" in mutual but not equal relations in nature-cultures (Haraway 2008a). Becoming in this co-evolutionary dance is "always

Figure 1.3 Lijnbaansgracht, Amsterdam (Photo: Katja Aglert 2000)

becoming with—in a contact zone where the outcome, where who is in the world, is at stake" (Haraway 2008a: 244). The potential openness of categories, and the plasticity of inclusive and excluding attitudes and actions towards other animals, poses specific challenges to sociology, challenges picked up by this book.

Urban Animals looks at human/dog and cat relations, with case studies concerning unleashed dogs, feral and homeless cats, crazy cat ladies, and urban animal hoarding. Cats, our most common pets, are characteristically defined in different ways based on both their relationship to humans (homeless, feral, domesticated) and the place they are moving in (domestic, indoor, farm, city). The cat's place is very much in the home (Grier 2006), but this is of course not universally the case. Needless to say, feral and street cats are ubiquitously present—for example, only in the city of Singapore, between 60,000 and 80,000 cats live in the streets (Davies 2011: 183). In many places around the globe, cats thrive in colonies, where neighbors care for them collectively (Miele forthcoming). The cat is also surrounded by a lush and historically constructed symbolism—unfaithful, deceitful, sensual, frivolous, but also enchanting and associated with the dark forces—and the magic lives on through history (Broberg 2004: 46). In fact, what contributes to our fascination with cats is probably that what characterizes them is their ability to embody any character (Rogers 2006). Thus, cats have been as tremendously popular in literature and visual culture as in real life.

Dogs historically inhabit several spaces in human culture, have the longest history of domestication, and are increasingly used both as companions and as working, hunting, and service dogs (McHugh 2004). In addition, they are used as experimental animals, religious objects, and food. Viewed as "man's best friend," dogs are popular figures in fiction, where they are often portrayed as loyal, loving, and faithful, but also as beastly and vicious (McHugh 2004). As a consequence of their close historical relationships with humans, dogs are also read in relation to their human companions, as cross-species identities (Sanders 1999). For example, service dogs are often portrayed as heroic, enabling their visually impaired humans to become civilly enabled—to gain human rights (McHugh 2011: 28). From a different perspective, many scholars have viewed pet dogs as degraded animals, deprived of any natural traits, and as such as pure products of modernity and urbanization (Tuan 2005). Sociologist Adrian Franklin (1999) explains the increased popularity of pets in the Western world with the notion that they fill the void caused by a lack of human relations in the individualized, modern city, thus contributing to—following Anthony Giddens' terminology—"ontological security." However, as Erica Fudge points out, it could similarly be argued that:

> pets by their very nature challenge some of the key boundaries by and in which we live and thus they cannot provide ontological security, but instead undermine it. [. . .] Maybe it is because the pet breaches a boundary, not despite it, that pets are so important in the modern world.
>
> (Fudge 2008: 19–20)

This is an intriguing possibility, and one can clearly see indications that pets, and particularly dogs, have become increasingly important in modern times (see Figure 1.4).

Dogs have also gone through an urbanization process, and, lately, urban places such as dog cafes, kindergartens, spas, cemeteries, and parks have been created. All over the Western world, dog parks as designated areas for dogs off-leash are now emerging, and these are often contested sites (Instone and Mee 2011). Increased dog ownership and changing pet relations, together with crowding processes also for non-human animals, lead to spatial conflicts.

Spaces, senses

Classical sociological accounts of the city, as developed by Georg Simmel and later by the Chicago School, conceptualized human experience as ambiguous (Judd and

Figure 1.4 Urban dogs, Bilbao, Spain (Photo: Tora Holmberg 2013)

Simpson 2011). Simmel (1972) approached the subject dialectically, and took seriously the tensions and dilemmas that he thought defined the modern experience. In his view, the process of becoming an individual includes both subjective interperson relations and more contextual, formal situations. Thus, individual and society, interaction and space, are in close proximity and cannot be separated other than analytically. The term "sociation" captures this dual process. To Simmel, sociation could be studied as an actual unfolding of social interaction in specific times and places (Levine 1972). For him, the study of sociation must involve a study of individuals' spatially specific experiences of social interaction in terms of "sensory perceptions," which are highly influential in social life (Simmel et al. 1997: 110). To Simmel, the city itself is always situated, with a possibility of transcending its own boundaries, through the community; politically, economically and "intellectually" (Simmel et al. 1997: 139). To be sure, Simmel did not consider sociation as something that could be shaped outside the purely human realm, working as he was within a humanist paradigm. Thus, I draw on Simmel's idea of subjectivity processes as ambiguous, involving sensory elements of conflict, repulsion, and aversion, along with desire, harmony, and sympathy, and read them through the idea of "co-evolution," "more-than-human," and "becoming with" as similarly ambiguous terms, with competitive and conflictual, as well as harmonious and caring, aspects. In connection to the affective dimensions mentioned (affection, repulsion), becoming with other humans in general and animals in particular has an important corporal dimension: to become is also to become a body and to make oneself available to the becomings of others, in what has been described as "anthropo-zoo-genetic" practice (Despret 2004). As in the case of the multi-species encounter between deer, dogs, and human described above, it is a matter of reading and understanding bodies across species lines, but also to move and affect bodies of other species, and to allow oneself to be moved and affected by another body. Moreover, when encountering other animals, verbal interaction is not primary (although some studies point at the importance of symbolic interaction; e.g. Sanders 1999). Instead, affect and sensorium are important embodied communication tools. Actors will use their vision, hearing, touching, and smelling capacities in order to create meaning and record and understand what is going on. Social science has recently witnessed what can be called an affective turn, particularly within cultural studies. Emotions, sensations, and affect become highlighted as objects of study, as well as research tools to help understand contemporary phenomena (Pink 2009). Animal studies have witnessed a similar trend, and scholars such as Ralph Acampora (2006), Josephine Donovan (2007), Vinciane Despret (2004), Donna Haraway (2008a), and Eva Hayward (2010) have sought to understand what affect and corporeality mean in terms of interspecies communication and the generation of more ethical relationships with other animals. Whether the answer to these questions reads "symphysis," "attention," "anthropo-zoo-genesis," "impression," or "response-ability," these authors and others carefully engage with both the philosophical and political implications of encountering other animals. Eva Hayward invents the trope of "fingeryeyes," accounting for the ways cup corals and scientists at a marine lab encounter one another "to name the synaesthetic quality of materialized sensation" (2010: 580). In her view, perceptions of other animals are generated by the interspecies encounters themselves. Moreover, she opens up for the interesting connection between species, senses and spaces, a connection I am obviously also interested in. According to Simmel and his sociology of the senses, "sensory impressions" cannot be reduced to something superficial in understanding social interaction, but:

> Rather, every sense delivers contributions characteristic of its individual nature for the construction of sociated experience; peculiarities of the social relationship correspond to the nuancing of its impressions; the prevalence of one or the other sense in the contact of individuals often provides the contact with a sociological nuance that could otherwise not be produced.
>
> (Simmel et al. 1997: 110)

Sociation is also always emplaced, thus attending to space as well as particular places is essential for urban sociology. For example, Louis Wirth writes about urbanism as a "way of life," indicating that urban space produces a certain kind of effect, of experience (1938; Tonkiss 2005). The city, for Wirth, is a "particular form of human association" (1938: 4), thus it is not strictly the number or the density of inhabitants but the organization of the social relations that defines the place as a city and conditions social life. Henri Lefebvre develops this idea and writes in *Urban Revolution* (2003 [1970]) of the dialectics of space and the everyday. For him the urban is less a set or objective reality, but a "horizon, an illuminating virtuality. It is the possible, defined by a direction, that moves toward the urban as a culmination of the journey" (2003 [1970]: 16–17). In line with Lefebvre rather than Wirth, I would like to address urban space as a process and a node for multi-species politics, rather than as a particular form. For me, this form is the city, the particular place; whereas space refers to the abstract dimension of the physical world, place is specific and contextual. Places are filled with meanings; they shape and are shaped by emotions (Casey 1993). Further, places are often contested terrains, where conflicting interests, notions and users struggle over access and definitions (Franzén 2002). Expanding on the human-centric notion of places most often utilised, Thomas van Doreen and Deborah Rose write about urban animals and their contribution to these controversial processes of meaning-making:

> Places are materialized as historical and meaningful, and no place is produced by a singular vision of how it is or might be. In short, places are co-constituted in processes of overlapping and entangled "storying" in which different participants may have very different ideas about where we have come from and where we are going.
>
> (van Doreen and Rose 2012: 2)

Storying, in their view, is interwoven with urban action and movement. I relate this to the work of urban sociologist Richard Sennett and his entanglement of bodies and places, of "social touching" and the built-in obstruction of the freedom to move (1994: 312). Only in relation to, and restricted by, others can we become our bodies. In fact, one could say that what is common to all animals is locomotion—the (potential) ability and will to move, to take action (Bull 2011)—and that this is what sociality relies on: "It is in locomotion, also, that the peculiar type of organization that we call 'social' develops" (Park 1967: 157). The movement of human and animal bodies in the city, and the politics around it, is a central theme developed in this book. In relation to this, I explore the figure of the moving stranger (or what Park (1928) calls "the marginal man"). In Simmel's terms, the stranger is the constitutive other, someone who is simultaneously fixed and unfixed in space, both close and distant from the group. The stranger may belong to a specific group, however, his/her status comes from the fact that the stranger brings along qualities that do not originate from the group. Moreover, the stranger has the potential to wander, to move away from the group, and is not bound by ties such as kinship (Simmel 1981: 149–151). Simmel also considered some degree of strangeness a dimension of all social interaction and relationships. This was later picked up by sociologists such as Erving Goffman (1990 [1971]) and Howard Becker (1963), who emphasize the dialectical construction of the norm and the deviant. Becker stated that: "[it] is likewise true that the questions of what rules are to be enforced, what behavior regarded as deviant, and which people labeled as outsiders must also be regarded as political" (1963: 7).

Thus, the construction of outsiders and strangers is a political, power-imprinted process, a conceptualization later successfully developed by Michel Foucault in relation to the perverted (1998), and by Sarah Ahmed as the racial other (2000). The ambiguity of the stranger and qualities of strangeness are fruitful to use in understanding the role of animals in the city—while constitutive of urban places, due to their status as belonging to nature, not completely belonging in urban milieus. The urban fox, for example, can change the ecology and create "passion" in various ways (Amin and Thrift 2002: 89–90). Similarly, the stray cat can be conceptualized as a stranger, as the temporal dimension of becoming feral, in different places depending on the meaning of that place (Chapter 3). However, the analysis does not stop at this point, but pushes the potential of transgressing these boundaries and furthers the notion of the stranger beyond the individual, in relation to humanimal crowding, as a possible arena of unruly and subversive relations.

Crowding

As indicated in its subtitle, this book investigates the co-presence of humans and animals in the city through the notion of "humanimal crowding." The notion is used as a "virtual object," something that by means of "transduction" will add to the more common methodology of induction and deduction (Lefebvre 2003 [1970]: 5). A virtual object can guide and perform analytical work as both a hypothesis about reality and as an object of inquiry. This comes close to what Haraway calls a figurative trope (1997), a material-semiotic knot in the web of science-fiction reality, potentially promiscuous in its origins, and open for both conservative and subversive uses and effects. Crowding (as well as the crowd), as social phenomenon and as a social scientific concept, has a heterogeneous history. Early studies of crowding were performed with rats, who were allowed to overpopulate in order for scientists to study the results in terms of behavior, morbidity, and mortality (Calhoun 1962). Later, these results were extrapolated to humans (Birke 2014). Thus, its multi-species origin and the "viral" component of crowding makes it even more fitting for my purposes (see Haraway 2012).

Closely related to the notion of crowding is that of the "crowd." The noun means a multitude of things, interesting for my focus on urban space and bodies but, first, "a large number of persons gathered so closely together as to press upon or impede each other; a throng, a dense multitude [. . .] the people who throng the streets and populous centres; the masses; the multitude" (*Oxford English Dictionary* 2012). A crowd in sociological literature refers to masses that could be either heterogeneous or homogenous, but often spontaneously created and conditioned by a shared identity or goal. Also common is the close connection between the emergence and control of a crowd and the formulation of a social problem. A more recent understanding of the term is that the crowd is defined by its members' proximity, both materially and figuratively (Drury and Stott 2011: 285). A long-lasting social scientific interest in the relationship between the individual and the collective gets tuned in on understandings of the crowd, seemingly displacing the distinction since it is more than a sum, and still not reducible to the individuals. Rather, the crowd often acts as one, although heterogeneous in terms of its parts. From the French Revolution onward, urban masses have caused unease, fear, and worry in both the ruling classes and among social scientists, who have formed alliances in order to prevent riots and control crowds (Sennett 1994). Andrea Mubi Brighenti suggests, in line with Deleuze, that the crowd can be viewed as something beyond the individual—not at a different level but rather as part of a different zone, called the molecular. This zone is the "undulatory domain of undifferentiated differences, where phenomena like crowds and packs occur" (2010a: 299). Crowds are thus modes, or ways of expression, rather than a number of aggregated individuals acting together for a certain goal. If they are undifferentiated, then according to a governmentality perspective, effective bio-political technologies must be able to transform the crowd, and thus reduce the multiplicities, into an identifiable unit (2010a: 300). Thus, surveillance techniques and practices have become popular, along with more traditional forms of population control.

To my knowledge, sociologists have not considered analyzing multi-species crowding and crowd control. So why am I interested in going down this path and reopening the notion of the crowd in this context? Within animal studies, as a contrast, the understanding of the collective has been of pivotal interest. Concepts such as swarm (Bennett 2010; Halberstam 2011) and pack (Deleuze and Guattari 2013) have been utilized in order to culturally understand a collective of non-humans/animals. Why do I not use any of these concepts instead? The reason for my choice is quite simple. I want to emphasize the sociological imagination since it so neatly captures the dialectics that I am interested in: individual/collective, subjectivity/form, action/politics, and experience/space. And, in the words of Haraway, the crowd may well involve the dimension of species:

> to be a species is to be constitutively a crowd, in symbiogenetic naturecultures, with no stopping point. Living piles turtles on turtles, all the way down. Species is about the dance joining kin and kind.
>
> (Haraway 2008b: xxiii)

In my understanding, the crowd is thus made up of, on the one hand, the undifferentiated collective and, on the other, the experiences and proximity of material-semiotic, human/animal encounters. However, the crowd is not explicitly a spatial sensitive concept. I therefore return to the twin concept of crowding.

The human urban population is increasing worldwide and, in Western countries at least, the number of companion animals is, too. Thus urbanization and its consequences are not purely human processes (Wolch 1998). Crowding as a phenomenon is well covered in urban studies, and originally it meant the increased density in population, due to urbanization (Churchman 1999). However, in the 1970s it became a concept that also captured the subjective experience, originally mainly stress and other negative consequences were highlighted (Stokols 1972; Gove et al. 1979). Nowadays, crowding is rarely used as an analytical concept, but essentially the same core is addressed when analyzing effects of urbanization. Micro-sociology has been employed in order to understand how people manage to live closely together, and the environmental, health-related, and psychological benefits of crowding in terms of high density have become more and more acknowledged. For example, in urban planning the terms "smart growth" and "compact cities" are used to highlight and implement the "good sides" of higher density. When it comes to official norms of proper housing, "crowdedness" as a concept has changed historically (Ekstam 2013), and the standards for measuring the lower limits of crowdedness are somewhat arbitrary (Ytrehus 2001). Many actors within the housing sector now choose to limit the number and breed of non-human inhabitants allowed. Crowding often resonates with the notion of vermin, and I will add crowding to other "verminizing" dimensions such as species, time, place, movement, and, not least, number. In fact, crowding may house all of these aspects. In the following, I ask how the figure of humanimal crowding—or the *spatial* formation of a trans-species collective in a certain social setting—plays out in the understanding and management of urban human/animal relations. I will in the following use the concepts crowd and crowding inclusively, the latter in order to highlight processes, while the former signifies a certain formation or product of crowding.

Zooethnographies

The study investigates urban controversies over companion animals, their understanding and management. Controversies are good to think with, since they expose otherwise hidden norms and values. Furthermore, norms and values may be strengthened, challenged, or subverted through the course of a controversy. For this purpose, I have conducted interviews with Swedish policemen and policewomen working with issues regarding animals, animal welfare inspectors and veterinarians from provincial offices, shelter workers, and activists involved in animal rescue, and people who have been subjected to official complaints (so-called hoarders). There is no specialty called animal police in Sweden, but the officers interviewed have the domain of human/animal relations as their main working task. Animal welfare officers in general have a number of duties, and their professional backgrounds vary from animal technician to veterinarian education. All of the interviewees are professionally well experienced with at least ten years in their field, some up to thirty years. All in all, twenty-one interviews have been conducted in four Swedish local counties ("län"), from the south to the north, in large cities and smaller ones. The interviews lasted between forty-five and seventy minutes, were transcribed, and then sent to the interviewees for modifications and approval. When quotes appear in the chapters, they have been made anonymous and translated into English. Included in the data are also texts and documents of various kinds: national legislation and policy documents, animal welfare protocols from inspections, and newspaper articles, along with international TV programs, reality shows, and social media. Thus, multi-sited, multi-varied materials are used in order to answer the questions posed. In addition, I have performed an ethnographically inspired case study of a suburban "dog beach" in the United States, analyzing the controversies surrounding it (Chapter 2).

Methodologically, I have been inspired by the ethnomethodological approach as developed by Harold Garfinkel (2002). Central to this approach is that people in their everyday lives—as members of certain social groups—try to make sense of the world through various methods, and that it is the aim of the social sciences to take this sense-making and the competences of the actors seriously (Garfinkel 1967). Thus, interpretive practices—the practices of sense-making—are of great concern to ethnomethodologists (Lynch 1993; Latour 2005). It is, in Anne Warfield Rawls' words, not so much a method, as an "attempt to preserve the 'incourseness' of social phenomena. It is a study of members' methods based on the theory that a careful attentiveness to the details of social phenomena will reveal social order" (Rawls 2002: 6). The ethnomethodological approach, as used in this book, seeks to understand participants' orientations and their productions of the social world through studying the ways in which members make use of their capacities. "Production" is itself a key term, since meaning-making and action are not separated, but intertwined dimensions (Francis and Hester 2004: 25). In a sense, this means studying culture in action, in becoming, through meaning-making strategies and resources (Hester and Eglin 1997). We make sense of urban human/animal relations through "interpretative repertoires" (Wetherell and Potter 1988: 171), in which bodies, senses, and spaces intersect. Interpretative repertoires are discursive methods of putting together certain terms and not others, in describing, understanding, and organizing the social world. I approach the data with the aim of understanding how human actors use such discursive resources in order to account for action, and to make themselves and others accountable in relation to cultural norms, and thus to create meaning. This *verstehen* approach does not exclude an analysis of social ordering. As noted by Rob Shields, a spatial analysis "reintroduces us to the complexities of the interplay between the different facets of social life" (1991: 10). Thus, an intersectional, ethnomethodological analysis adds to an understanding of the everyday life and the spatial production of meaning.

Of course, one may rightfully ask how an essentially humanist ethnomethodological approach can be combined with a study of human/animal relations? There are naturally some traps and flaws inherent in the anthropocentric attunement to discourse studies and analyses. I could, for example have chosen to include ethological perspectives on animal behavior. However, I chose not to, because I want to focus on the becoming in human/animal encounters, and consequently not essentialize animal behavior, while keeping a constructivist approach to human performances. In addition, I could have elected to study in more detail the interaction between individuals of different species (Sanders 1999). However, as I became more and more occupied by humanimal encounters beyond the individual, this approach was no longer desirable. To this end, I have been influenced by the "zoo-ethnographies" approach that has developed through discussions in the Humanimal group in Uppsala. Through an international symposium on the matter in 2011, as well as in a joint publication (Humanimal group forthcoming), we have problematized the lack of focused discussions around methodology in animal studies. While not aiming for a homogenous approach, we want to use the "humanimal lens" in order to question methodological conventions: "Consequentially a zoo-ethnographies approach calls us to examine the methods and methodologies implemented in our zoo-sensitive enquiries, to critique methodological orthodoxies and use the existing approaches and methods creatively" (Humanimal Group forthcoming). "Zoo-sensitive enquiries" could obviously mean a range of things, challenging existing methodologies and writing conventions—some breaking with the boundaries of science/arts (Snaerbjörnsdottir and Wilson 2011), some creative in research design (Driessen et al. 2010), some in academic prose (Hayward 2008), and some in the framing of research questions (Bull 2014). For the present study, this approach means keeping a critical eye on the anthropocentrism inherent in interviews and text analysis, as well as in the representations of animals in the book, through the practice of writing "nearby" animals. Writing nearby (cf. Chen 1992), as a methodological standpoint, means that although I will no doubt fail to engage and represent the animals in this study, I will nevertheless try by focusing on the relationality of humans and other animals. Because—and this is my last methodological remark—if humanism is about the rejection of everything non-human, more-than-human is not about excluding humans or human experiences, but de-centering them in favor of species relationality. In line with this standpoint, I will not throw the human out with the bath water, but keep the empirical focus on human action and experience, while the humanimal relations constitute the endpoint of the analyses.

Outline of the chapters

In order to investigate the questions and issues raised so far, four empirical/analytical chapters will lay the foundation for the conclusions and discussions in the last two chapters of the book. Part I, "Animals in the city," starts with Chapter 2, "Bodies on the beach: Allowability and the politics of place," and investigates the emergence and continuation of a spatial conflict concerning a "dog beach" in Santa Cruz, California. The case study is used as an example of how urban politics affects the bodies, practices, and movement of people, dogs, and other species, but similarly how these politics are constantly under threat due to civil disobedience and subversive acts of counter-politics. In particular, the concept of "allowability"—sociospatially produced inclusion and exclusion—is developed as a dimension of the process of politics of place. The chapter discusses how the collective movement of dogs and people in (sub)urban space can be conceptualized as a trans-species urban crowd, threatening a certain public order. Chapter 3, "Stranger cats: Homelessness and ferality in the city," discusses a specific area of urban controversy in Sweden, namely the understanding and management of homeless and feral cats. How are "feral," "homeless," and other categories produced and with what consequences? What is the role of "home" and other places in defining and handling various categories of urban cats? The categories "lost," "homeless," and "feral" are defined and made meaningful in the context of the juridical regulations, the actors involved and, not least, the site of the action. Consequently, places play important roles in accounting for meaning-making and the management of homeless and feral cats, where the interplay between home-ability and ferality is the decisive methodology of "cat-egorization." The stranger cat embodies a strangeness that, through its ambiguity, carries the potential of radically questioning the meaning of places such as home and street. However, in law and urban homelessness management, potentially heterogeneous meanings get reduced to being a body out of place, a threat to the neighborhood as well as to human health, and normalized through crowd control technologies such as adoption practices, trap–neuter–release, or euthanasia.

Part II, "Humanimal transgressions," consists of two chapters, of which Chapter 4, "Verminizing: Making sense of urban animal hoarding," is the first. The chapter investigates the phenomenon of so-called animal hoarding, the construction of deviance, and the effect of stigmatization processes—the strengthening of norms—by examining explanations and management of urban animal hoarding. Behavioral scientific accounts, Swedish animal welfare data, interviews with so-called animal hoarders, and analyses of representations of hoarding in media reports and US reality shows, are used to answer the questions of the chapter. Towards the end, the analyses are connected to the broader framework of humanimal crowding and it is argued that hoarding can be viewed as a verminizing phenomenon. Through normative notions of the urban home, humanness, and animality, I discuss how hoarded crowds, and the meaning they take through the spatial context, can be understood through the lens of trans-species sensuous emplacement. Chapter 5, "Feline femininity: Emplacing cat ladies," furthers the analyses of animal hoarding, and explores representations of "crazy cat ladies" through highlighting the intersection of e.g. species and gender. Women devoted to cats are often described as eccentric and in need of an explanation—for example, early experiences of deprivation or abuse, infertility, sentimentality, and femininity. In this chapter I ask how cat ladies are portrayed in international popular culture and within urban management. The analysis of these images highlights the prevailing norms of emplaced gender and human/animal relationships. Through an intersectional approach, a kind of subversive performance—that of "feline femininity"—is outlined.

Part III, "The promises of crowding in zoocities," starts with Chapter 6, "Beyond crowd control." This chapter revisits the aims and research questions posed in the introduction, and draws together the findings of the different studies. Departing from the findings of the book, it shows that the multi-species city includes opportunities as well as constraints for human/animal relations. The exclusion of unruliness may be an urban ideology, aiming at social order, but close attention to the level of practice reveals a much more diverse, disordered, and perhaps disturbing experience. But what could this potentially mean in terms of multi-species futures, urban ecologies, human/animal ethics, and the embracing of heterogeneity? Bringing "humanimal crowding" to the fore once more and highlighting the promises of the crowd—humanimal proximity and sociation—it is set in conversation with more-than-human urban politics as discussed through, among others, Jennifer Wolch's framework of "zoöpolis," Steve Hinchliffe and Sarah Whatmore's "politics of conviviality," and Chantal Mouffe's "agonistic pluralism," in order to point to an alternative mode of more-than-human politics: one that takes seriously the ubiquitous and potentially transformational relation of body and the city, action and politics. In Chapter 7, "Open endings," I discuss—in a conversation with Stockholm/Berlin-based artist Katja Aglert—the findings presented in the book using a less chronological, chapter by chapter, approach. Instead, she and I dwell on some themes that run through the book as a whole: boundaries and liminality, sensuous flows, transformations of individuals and collectives through crowding, verminizing, messiness and relationality, as well as the epistemologies of human/animal encounters. What the chapter brings out is a kind of wonder concerning the connectedness of the specific human/animal encounters analyzed throughout the book, using a much broader ontology of modernity, purification, taxonomies, and identity politics. The chapter also suggests and embraces an analytical kinship to seemingly different urban phenomena: the handling of feces and other waste through urban metabolism, subcultural formations such as graffiti writing, along with public manifestations and social movements.

Part I
Animals in the city

2 Bodies on the beach

Allowability and the politics of place

This dog is a Labradoodle. Half poodle, half Labrador retriever, like most of the locals she too could claim French-Canadian roots. Either way, I'd like to see this dog like me as "from away," as they say here of everyone not born of Mainers. I'd like to think she shares my suspicion of their shifting notions of what's allowed. Worse, I've fast-talked and fast-walked her around the law for too long to stop now. On the beach, we see no fences, no markers, so we just keep going.

(McHugh 2010)

This chapter concerns the issue of urban human/dog politics: where, when, and how are dogs and their companions allowed to share the urban space, and how are they produced due to processes of inclusion/exclusion? The domestication of dogs, a process with roots going back 10,000 years, is still ongoing and unruly (Cassidy 2007: 20). It is arguably a relational process, by which both humans and dogs are domesticating one another (Haraway 2003). Moreover, domestication is a spatial process, developed under certain conditions, in certain places, by certain actors and through particular struggles (Andersson 1997, 1998). In the West, urbanization and a general hygienic discourse have led to increased exclusion of, and prohibitions concerning, dogs. In contemporary cities, dogs must often be leashed at all times, even in parks. Cities like Vienna and Berlin stipulate that large dogs must wear a muzzle in public (although, in practice, they seldom do). These prohibitions would seem to suggest that dogs do not belong to the civilized city—unless they behave properly. There are of course exceptions. In Sweden, for example, dogs can ride on public transport, and in many southern European cities, it is rather rare to see dogs on a leash. To complicate things a bit further, people and dogs do not always follow the rules, but may challenge norms and regulations.

Moreover, a contrary trend is operating to include dogs and their people through a number of designated, canine-specific places: dog parks, dog cafes, dog salons. While urban space in general is becoming more purified, dogs and their owners seem to be carving out some spaces, both by breaking rules and by sidestepping them through the creation of new, "quasi-exclusionary" places (Tissot 2011: 265). These new movements have interesting effects in terms of

Figure 2.1 Prohibition sign, Santa Cruz (Photo: Tora Holmberg 2010)

class, ethnicity, and gender. Sylvie Tissot (2011) discusses how the creation of a dog park in an inner-city area in North America helped gentrifiers hold on to a discourse of diversity, while at the same time excluding—disallowing—certain undesired clientele. At a more mundane interactional level, other studies demonstrate how regular dog park visitors guard the place from newcomers, socializing and monitoring human and dog behavior for weeks, before including them as legitimate visitors, as members (Robins et al. 1991). Others show how such dog parks often are placed on unplanned land, for example on the sprawled outskirts of cities, in industrial neighborhoods, or in less pristine areas of city parks (Instone and Mee 2011). How can we conceptualize these seemingly contrary trends of inclusion and exclusion of the dog/human node? Drawing on Foucault's twin power regimes, Philip Howell suggests that, for at least a century, muzzling and quarantine—disciplinary technologies to prevent dog bites and rabies—have co-existed with, but gradually been replaced by, governmentality-oriented technologies including leash and voice control (Howell 2012: 222; see also Michael 2000).

Thus, we have recently been witnessing a process of domestication through the "model of responsibility," one that allows dogs and their walkers to enter public space if they govern their conduct and behave properly (Howell 2012: 235).

> What we have here is something like the domestication of public space. Just as animals were welcomed into the private space of the home, in the form of the "pet," so were animals allowed to be properly public in a domesticated and liberal public space. [...] But we may go a little further still, however, and make space for the dog not just as a sort of liberal political subject, merely allowed into public space but perhaps as an agent in its own right.
>
> (Howell 2012: 240)

Through domestication processes similar to when the dog was turned into a pet through its inclusion into the private home a century ago, dogs and their companions in urban space have now been "conditionally domesticated." In this chapter, conditional domestication is understood in terms of "allowability"—or the sociospatial production of inclusion/exclusion and the process of becoming admissible—which calls for critical questions: where, when, how and why are the dog/human nexus allowed or banned? The heart of the problem lies in understanding how the agency of dogs, people, and other kinds play out in particular urban places. In this chapter, I seek to understand the "politics of place" in

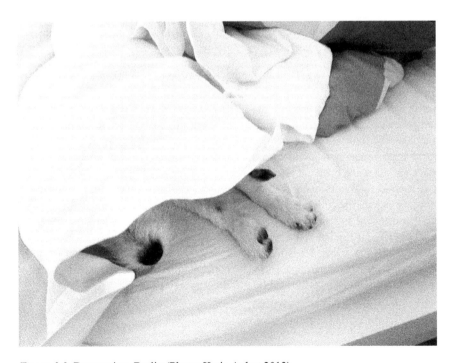

Figure 2.2 Dog resting, Berlin (Photo: Katja Aglert 2013)

Figure 2.3 Dog resting, Thailand (Photo: Katja Aglert 2013)

action, departing from a specific place, Its Beach in Santa Cruz, California. How do processes of inclusion and exclusion play out in action? How do actors understand their actions and make them accountable? In a more general sense, however, the chapter concerns our understanding and construction of the concept of urban space and its place in social interaction, and how it can be reconceptualized when it is rubbed up against and forced to accommodate the species dimension. By bringing together sociological classics, spatial politics research, and animal geographies oriented toward the nature/culture divide, I intend to demonstrate how everyday life is affected by politics, but also how it constitutes a site for resistance (Lefebvre 1996). Thus, I wish to discuss how the collective movement of dogs and people in (sub)urban space can be conceptualized as a trans-species urban crowd that threatens a certain public order.

Urban space revisited

Writing in the early 1900s, Simmel critiqued his precursors for neglecting the uniqueness of spatial qualities, for reducing the spatial to mere physical form or a context for social interaction. Instead, he wished to qualify our understanding of space and suggested a number of important dimensions, or "fundamental qualities of the spatial form upon which the structuring of communal life relies" (Simmel et al. 1997: 138). These qualities or dimensions are as close to methodological cues as one gets in the works of Simmel. Because they are of importance for the arguments I make in this chapter and the next, I will look at these dimensions in some detail.

The first is the exclusivity of space, by which Simmel means that each and every space is unique, even though different spaces may be analogous. Moreover, the uniqueness of the space forms the objects within it, such that the same object will be different depending on where it is spatially situated. Spatial ordering becomes obvious when we think about urban sociality. An encounter with a cow at a zoo is a completely different experience than encountering cattle in the shadow of a skyscraper (Sabloff 2001: 13). Sociologically, this means that all social encounters need to be attended to in the making, within their spatial and temporal context (Garfinkel 1967). The social is always already spatial, an idea taken up by several urban theorists, most famously by Lefebvre, who more directly than Simmel attends to physical space and objects, attributing to them social status (Lefebvre 1991). Similarly, a space is defined and changed by the objects, actors, meanings, and emotions taking place, and, dialectically, framing the interaction going on.

The second dimension is the compartmentalization of space. As an example, the urban zoo is structured for certain forms of interaction through the architecture of walking, looking, and encountering. Thus, species-imposed order and ordering is the norm. However, even the strictest order has dimensions of messiness. As stated in Chapter 1, the noun "zoo" signifies both chaos and order. In practice, space is divided into sections, separated and socially framed by boundaries, "both as a cause and an effect of the division" (Simmel et al. 1997: 141). For the argument of this book, the significance of this observation, made more than a century ago, lies in our understanding of the contingency of spatial boundaries. Simmel states that natural boundaries are not decisive. On the contrary, the spatial demarcations are arbitrary with regard to materiality. Cities, neighborhoods, parks, streets, islands, and beaches are constructed—or in Haraway's terms, material-semiotic figurations—and, as such, produce their own meaning. A park requires certain codes of conduct, set apart from the rules applied to a street (Borden 2001; Kidder 2011). As an illustrative example, Sharon Zukin talks about the methodological difficulties in doing comparative fieldwork on local shopping streets. What may look like a street to one person appears to be a block to another, depending on the cultural heritage and the social, commercial, and residential history (Zukin et al. forthcoming). This dialectics of form and interaction is emphasized in Simmel's sociology of space. Moreover, the park derives its meaning in relation to the street, so that every unit "gains its spatial expression in the enclosing boundary" (Simmel et al. 1997: 141).

But boundaries are potentially controversial; as liminal spaces they embody a tension. Simmel exemplifies this tension with the boundary drawn between two neighbors, saying that it is at once offensive and defensive. He is clear, however, that it is not the boundaries themselves that are in tension, but the actors involved in that space—inhabitants, proprietors, strangers—who through their action both produce and reproduce boundaries, and signify their meaning. Every boundary is, in his view, a sociological event. This theme has been picked up on by sociologists from various fields and specializations, including historical sociology (Elias and Scotson 1965), science and technology studies (Bowker and Star 2000), cultural sociology (Bourdieu 2001), and animal sociology (Hobson-West 2007). The question remains, "How do animals contribute to the production of space through the makings of boundaries?"

The third dimension of interest is the capacity of space to fix its social formations. Simply put, to what degree are the social relations determined by the space and its objects? To what extent can social relations extend the spatial boundaries? Simmel exemplifies this dimension of spatial quality by contrasting the insurance market with the rendezvous, where the former is only loosely bounded, while the latter signifies both the nature of the encounter and its location. "The sociological essence of the rendezvous lies in the tension between the punctuality and fleeting quality of the relationship, on the one hand, and its temporal and spatial determinacy, on the other" (Simmel et al. 1997: 148). This is not to say that the spatial context determines the relation but, once again, it is the dialectics between space and interaction that is highlighted. The spatial influence on social formation is closely related to the issue of proximity and distance, and how the margin is conceptualized (Shields 1991: 3ff.). The spatial quality of interaction between individuals as well as organizations, if they are in contact or separated from one another, generates commonality in terms of interests, meaning making, or forces. And the reverse is equally true—a strong commonality creates close interaction (Elias and Scotson 1965). Simmel adds to this sociological line of argument, first by emphasizing the relativity of concepts such as proximity and distance, and, second, by highlighting the importance of feelings—of sensuous encounters.

> Alongside the obvious practical effects of spatial proximity and the sociologically highly important consciousness of having those effects available at all times even if one has no desire to make use of them at the moment, the result of proximity for the form of sociation is composed of the significance of the individual senses with which individuals mutually perceive one another.
>
> (Simmel et al. 1997: 155)

This quote densifies the argument that the degree of spatial proximity and its impact on "sociation"—that is, the socio-spatial formation of interaction—are sensuously negotiated. One thing needs to be clarified, namely the meaning of "individual senses" in the quote above. To Simmel, senses are strictly social, however individually experienced and expressed they may be. Emotions bring people together, or lead to repulsion. Sarah Ahmed (2004), among others, has elaborated on this issue, and speaks of emotional spaces as culturally produced.

Summarizing this long excursion into Simmel-land, there are some important questions that I wish to bring to the theoretical discussion on multi-species urban space in this book, as well as to the analysis of the case study explored in this chapter. What are the unique qualities of the space studied, and how do these qualities frame the objects and subjects within it? How are the boundaries of this space understood, and how do these understandings frame a certain social ordering? To what extent are the contested qualities of the space fixed or transcended in time and space? And how is the proximity/distance dimension negotiated sensuously, and with what consequences? The next step is to move from the abstract space to the more concrete place, and the practices and identity politics surrounding it, and then to ask whether and how the dimension of species can challenge the proposed scheme—a task for the discussion below.

Politics of place

In this chapter, I use an ongoing debate to uncover more-than-human politics in action. Debates are fruitful contexts for capturing unspoken norms and values, along with taken-for-granted practices (Garfinkel 1967). Norms, ideas, and identities are produced and reproduced through debate (Mulkay 1997). In this case, it is more precisely the "politics of place" that is the context, meaning historical-cultural struggles to construct, define, and settle the hegemonic meaning and function of a particular place—the question being who has the right to access, to sensing who belongs and who should be excluded (Franzén 2002). With regard to a square in central Stockholm, Franzén writes:

> As a place, Sergels torg is marked by the uncertainties and ambivalences of modernity, while both sides in this politics of place are seeking closure, trying to make the square into something unambiguous—either a cosy place, or a historical monument.
>
> (2002: 1123)

Similarly, it can be said that the controversy over Its Beach, Santa Cruz, presented in this chapter, is one in which actors struggle to define and establish a definite meaning, a "sense of place" (Feld and Basso 1996). Thus, here I will recount some recognizable features of contested everyday socio-spatial relations, taken from, among other sources, studies of skateboarding (Stratford 2002; Chiu 2009), graffiti writing (Brighenti 2010b), young street dwellers (Valentine 2004), and punks (Hebdige 1979). In the literature on human/animal relations, similar politics of place and struggles over definition and historically situated access and practice occur in contexts such as urban pigeon feeding in London and Venice (Jerolmack 2013), bull racing in Andalusia (Thompson 2007), flying foxes in Sydney (van Dooren and Rose 2012), and elks in a park in Gothenburg (Holmberg 2011a). Challenging urban ecologists, the social interaction of humans and animals in the built environment is not just a surrounding, but is embedded in meaning-making practices (Jerolmack 2013: 75).

In sociological micro-studies of spatial conflicts, including those involving human/animal relations, meaning-making processes and discursive struggles are highlighted in great detail. Paradoxically, however, what I often find missing is the spatial dimension—as understood by urban classics—the physicality of places. "Whether built or just come upon, artificial or natural, streets and doors or rocks and trees, place is stuff. It is a compilation of things or objects at a particular spot in the universe" (Gieryn 2000: 465).

As referred to above, place structures and produces social relations, and vice versa. And the process of sociation, of becoming one through the dialectics of form and interaction, deeply involves bodies and emotions. Members are not "in" a place, but produce it through various bodily adjustments and practices. They are emplaced. A playground is produced through play, but could be contested through other bodily and discursive practices such as marijuana smoking and public drinking. In the playground example, people as well as other animals make use of the carousels and slides to sit on, or the huts to hide in. Children may use the playground in the daytime, and leave it open to others to use the area for different purposes at night. The playground will still add a certain playfulness to the crowd, as the practice of smoking and drinking in public may be produced as childish, even innocent. Thus the spatial conceptions or "place images" produce and structure everyday life (Shields 1991: 6). Places that allow for such appropriation and use in unintended ways can be conceptualized as "loose spaces" (Franck and Stevens 2006). Loose spaces are needed in cities, because they promote diversity, vitality, and urban renewal. However, the use of, for example, the playground for other purposes and by other actors than was originally intended may be thought of as illegitimate, as an intrusion on the children's space, and may lead to controversies. A Canadian study of the conflict between dogs, their companions, cyclists, and children illustrates that dogs are conditionally allowed (Patterson 2002). It is all a matter of where and when, and in the company of whom. Thus, animals, humans, objects, materials, emotions, scale, and space, "assemble" the place through enactments, through action (Farias and Bender 2011: 6).

Extending this line of argument, Colin Jerolmack notes that, in the struggle over definitions and uses of the emblematic place Trafalgar Square in London, pigeons, feed vendors, tourists, animal activists, security guards, and hawks act so as to produce contrasting versions of the place. Proponents of the square as one for pigeons and their feeders have made innovative use of the space to bypass legislation and continue their feed exchange and other interactions. In Venice, concrete actions to limit pigeon presence include a recent prohibition against feed vendors, and architectural innovations such as nets and spikes. Jerolmack wittingly notes that, "Pigeons have become particularly despised urban trespassers because they, in all their animality, are so public. It is almost as if they taunt us with their seemingly 'unnatural' predilection for stone and concrete" (2013: 73).

But these prohibitions seem to do little to limit the pigeon presence. Increased regulations only seem to produce new practices, such that the "enchantment" of the pigeons in producing the meaning of certain places, including the built environment, continues (Jerolmack 2013: 75).

Urban scholars have recently investigated the dialectics of form and subjectivity, including the appropriation of material structures (Borden 2001). Following

Figure 2.4 Anti-homeless spikes in London (Photo: Andrew Horton 2014)

bike messengers and their playful work in the streets of large North American cities, Jeffrey Kidder contends that the material environment is produced in and through social practices (2011). Kidder's bike messengers use and manipulate the uniqueness of the space, which in turn generates the "lived emotions of messenger practice"—a corporeal and situated, emplaced activity the author calls "affective appropriation of space" (2012: 13). Building on Simmel and inspired by similar recent developments in urban studies of Lefebvre's notion of "lived space" as the appropriation of spatial forms, including the physical milieu, I discuss politics of place through trans-species corporeal interaction and spatial formation.

Dogs and liminal space

As mentioned in the introductory chapter, dogs and other pets can be understood as liminal—to us, they are both human companions and "other" (Fudge 2008; Redmalm 2013). Thus, inclusions and exclusions, and the drawing of boundaries through binaries, are of great interest to study, as they will reveal the cultural deep structures of meaning and ordering (Hannerz 2013). In my case, I am particularly interested in the dialectical experiences of such orderings of liminality.

Let us now zoom in on the liminal space and place under study here: the beach. In *Places on the Margin* (1991), Rob Shields writes about the shifting and contradictory "social spatiation" of places, which includes time, space and social structuration dimensions. For analytical purposes, he introduces the term "place images." Place images, he writes, are neither historically stable facts nor fiction waiting to be filled with meaning. Instead they are "produced historically, and are actively contested" (Shields 1991: 18). In Shield's analysis of social spatiation through images of Brighton, England, he argues that the town, as a seaside resort and throughout history, rests on the notion of liminality.

> Its shifting nature between high and low tide, and as a consequence the absence of private property, contributes to the unterritorialised status of the beach, unincorporated into the system of controlled, civilized spaces. As a physical threshold, a limen, the beach has been difficult to dominate, providing the basis for its "outsider" position with regard to areas harnessed for rational production and the possibility of its being appropriated and territorialised as socially marginal.
>
> (Shields 1991: 84)

This liminality, in turn, produces and is produced by a certain carnivalesque atmosphere, an escape from the temporally and socially structured Victorian life, and an "occasion for the enactment of alternative, utopian social arrangements" (1991: 91). Thus, again it is the dialectics of space/subjectivity, of form/action, that is elaborated, here in relation to liminality.

Cultural studies scholar John Fiske (1989) reads the beach as a cultural text, uncovering the ideological power structures (race, class, gender) that are spelled out at a particular beach. He also notices some general features of the urban/suburban beach, as opposed to the "wild" beach. According to Fiske, the beach is an anomaly between nature and culture, where certain parts (lawns, parking lots, toilets) belong more to culture, while water along with waterfront (diving and swimming) is more a part of nature (1989). In our cultural imaginary, the sea is often perceived of as more "nature" than anything else, and is often constructed as the last frontier: "Provoking not only ideas of margins, exchange and openness, the beach, particularly in an Australian context, also represents a deeply contested site, suggesting struggles over issues of ownership, belonging, nationality and culture" (Brown et al. 2007). Moving from Australia to the United States, I will now turn to a particular town in order to study the more-than-human politics of place in action.

A history of a beach

Santa Cruz is located 115 km south of San Francisco in central California, with a population of close to 60,000. This small Californian city is on the stricter side in the US when it comes to dog politics, as regards restricting and regulating the movements of and spaces allowed for dogs. The city invests great resources in compelling dogs and their people to abide by the law. However, it is not surprising that, although dogs are prohibited altogether in certain areas and dogs off leash are banned—and consequently there are some serious risks and punishments associated with breaking the ban—rules are continuously disobeyed; spatial restrictions only produce new forms of trespassing (Borden 2001: 254). This is the context of a controversy regarding an unofficial urban/suburban "dog beach," a place that people from different interest groups struggle to gain access to and to define using all possible means. The case study will be analyzed as an example of a politics of place, in which several actors are trying to define and enact a certain location. The question is, in brief, "Is this a beach for off-leash dogs or not?" However, the case offers an additional complication, namely the ambiguity of human/dog relations in the urban space. It is used as an example of the dialectics of everyday lives—of the bodies, practices, and movement of people and dogs—and space: the liminal case of the beach.

The data presented come from fieldwork performed at Its Beach (belonging to the Lighthouse Field Park, by West Cliff Drive) during three weeks in April, 2010. I had heard of a conflict regarding the beach, and set off to observe and talk to people about their experiences of going there. Why did they go there? How did they feel about the place? How did the dogs and people get along? Did they experience any conflicts? When did these occur? Had they encountered the police or other authorities during their visits? These ethnographic accounts were then used as a means to start unfolding the controversy, through websites from involved neighborhood organizations, official documents from the State of California and the Santa Cruz City Council, and blogs from people visiting the beach. The method is thus not a traditional ethnographic one, but inspired by the approach of "following the controversy" (Latour 1988).

Santa Cruz, situated on the central California coastline, on the northern tip of Monterey Bay, is predominantly a white, middle-class city, where the percentage

Figure 2.5 Dogs on beach, Barry Island, UK (Photo: Joao Bento 2013)

of the population consisting of other "races" is significantly below the state average, while educational background shows the reverse pattern; the percentage of the population holding a bachelor's degree or higher is above the state average (City Data 2010). The beach of interest for the present study can be reached from West Cliff Drive, flanked by the large villas and bungalows of the Lighthouse neighborhood. This 0.518 square mile neighborhood is a solid one, with a median annual household income in 2008 of $72,195, compared to the Santa Cruz median of $63,227, and the houses are significantly larger than average (City Data 2010). The Lighthouse neighborhood surrounds the Lighthouse Field State Park, and the connecting Its Beach. It is a popular park and beach, enjoyed both by humans—not least tourists and surfers—and by sea lions, birds (such as the rare black swift), and dogs.

Almost 40 percent of all Californian households have a dog, but less than two percent of state and city parks are off-leash areas (Dog Park USA 2010). In Santa Cruz, there are no fenced dog parks, but a few parks and one beach allow dogs to be off leash during certain hours. Until recently, the Lighthouse Field Park and adjacent Its Beach were open for such off-leash recreation (between sunrise and 10 am, and between 4 pm and sunset), much due to the area's status as an undeveloped state park. The local interest organization, Friends of the Lighthouse Field (FOLF), writes in a petition to state park director of a "50+ year tradition of voice control/off leash recreation," but also that "the beach portion (Its Beach) has always been heavily used by the public, not only by people with dogs but also surfers, runners, sunbathers, dance and drummer circles, and many other recreational users" (FOLF 2010). In May 2002, some of the dog-accompanying visitors started to sense a threat to the tradition, and formed the FOLF group, with the aim to "support the preservation of the beauty and recreation opportunities for people and dogs off leash at Lighthouse Field and Its Beach" (FOLF 2005). All the neighbors and other visitors to the park were not particularly happy about the presence of dogs running around and, in connection with a plan update (lasting from 2001 to 2003), a group of people—calling themselves Lighthouse Field Beach Rescue—complained to the officials, asking for new and stricter regulations. Later, in 2003, a lawsuit was filed from the same group, demanding an environmental investigation regarding the canine impact in the area. In 2005, this lawsuit was partly approved, when the Santa Cruz Superior Court ruled that, even though there was no environmental study of the effects of unleashed dogs, there was not enough evidence to force preparation of an environmental impact report for the park plan: "Once the informational requirements of a complete initial study have been met, the city as lead agency may again determine whether a negative declaration, a mitigated negative declaration or an EIR is appropriate" (Superior Court of California 2005).

The plan update controversy almost exclusively regarded off-leash dogs, and there were numerous public hearings, environmental studies (looking at, for example, the effect of dog feces on water quality and the effect of dogs on wildlife), petitions, and surveys in the years between 2001 and 2005 (however, the controversy is still ongoing). After the court ruling in 2005, the city council finally decided to drop the update, which meant leaving the rules unchanged. Then suddenly, in October 2005, the State of California overruled the city's decision, and stated that "continued off-leash dog use of Lighthouse Field State Beach cannot be reconciled with the statute's current mandate that any animal brought into a state park must . . . be under the immediate control of the visitor or shall be confined" (Torgan 2005). One consequence was that the park (and the connecting Its Beach) now fell under the same unconditional off-leash ban as the rest of the state's parks, a ban that came into force in November 2007. Now, how did the parties argue for and against in this case—in short, how did they account for the different versions of the beach performed?

Safety/risk

One strong theme evolved around the issue of safety versus risk. FOLF, on the one hand, states that the presence of dogs and their people at all times of the day make the Lighthouse Field Park a safer place for everyone. This is contrasted by the current situation, in which the park is either not used at all, or is used by homeless people and drug users. Thus, the latter are framed as illegitimate users, while dogs and their owners are produced as legitimate. On FOLF's web page, we can also read that off-leash exercise leads to more "socialized" dogs that are less likely to be aggressive in other environments (FOLF 2010). In essence, dogs who are trained in being around other dogs learn how to behave in a pack. Overall, it is stated, the park and the beach is a very safe place for people and their dogs, and only one incident of a dog bite has ever been reported. This is confirmed by a Santa Cruz Animal Services Field Supervisor, who in a hearing arranged by FOLF in 2007, stated:

> I've been working here for four years, and I would say no. I've been responding to that location for dogs falling off rocks, being in areas where they are not supposed to be, falling from cliffs. [. . .] But no. I know about three years ago, I did a lot of patrol in that area, out of uniform, trying to catch people not picking up their [sic] defecation and in that time I probably logged about 80 hours of controls, and only caught one person not picking up defecation, so . . .
>
> (Stosuy 2007)

The hearing and its testimonies are available through YouTube and, together with other PR activities, demonstrate the cultural resources of this group; they know how to mobilize public awareness and they use a range of available channels—T-shirts, stickers, petitions/letters/phone calls to officials, internet broadcasting, and local campaigns—in order to assemble the beach as a safe place for everyone. On the other hand, there are complaints about the threat of dogs to human safety, and thus, a call to closer control of their whereabouts:

> I have been chased and bitten by off-leash dogs while minding my own business. I don't care to ever encounter dogs off-leash. [. . .] I support the

creation of dog parks away from where I might encounter them. I loathe irresponsible dog owners and I vote.

(Rescue Santa Cruz Beaches 2004)

In this statement, the speaker presents himself as a good citizen, someone who, in contrast to the "irresponsible dog owners," is responsible, minds his own business, and votes. Note how the author wishes the dogs, or perhaps rather the opponents, were somewhere else, a common rhetorical strategy in this and other urban, spatial conflicts (Elias and Scotson 1965; Weszkalnys 2007) and one used by both sides of this debate. Through the accounting of unleashed dogs as unsafe, and their owners as "loathable," they are produced as lacking allowability in this ongoing politics of belonging.

It is worth noting that there is no clear-cut divide between dog owners and non-dog owners when it comes to taking sides in the debate. In fact, some people testify that they are "good" dog owners to make a stronger case for a ban:

I am a dog owner who feels strongly about keeping my dog on leash for both her protection and the protection of the environment. I let her run free only when I am certain it is safe. I am in favor of a fenced dog park where all of us who go there understand it is a dog's place to roam.

(Rescue Santa Cruz Beaches 2004)

Here, safety is a key sign, "protection" in the name of the dog and the environment, and off leash only "when I am certain it is safe," meaning that the area is fenced. Physical boundaries and activity-specific rules, or rather the lack of such, are produced as unsafe, both for dogs and humans.

Risk and safety are interesting dichotomies in many regards. First, what is perceived as risky is clearly a contested matter. While one person sees dogs on the run as a threat to personal and bodily safety, another views them as a guarantee of communal safety, both in the park and elsewhere (dogs become less aggressive). Second, the genre of personal testimony is used to argue for more safety. While one has been bitten, another states that there have never been any incidents in the area. Clearly, the past is expected to determine the future. However, because there is no agreement on the nature of past events, the future is likewise a contested site (see Brown 2003). Another common denominator is that safety—regardless of its definition—is produced as desirable, a contrast to risk. The beach should be a safe place, not a risky one. But, there are more actors enrolled in the production of risk.

Disturbance

One of the arguments for banning (unleashed) dogs in the area is the concern for the so-called natural resources. Parties agree that the area is the habitat for a number of protected or otherwise special species that people should be concerned about. But, if and how the presence of dogs disturbs such wildlife and the extent of the protectionist agenda are debated issues. For example, in a petition filed to the Superior Court of California in May 2003, it was stated that the environmental studies performed by the City of Santa Cruz were insufficient, as they did not take into account with sufficient seriousness how the—in the complainers' view—increased presence of dogs would have an impact on wildlife.

> Among other environmental issues raised were the inconsistency of the Plan with resource and wildlife management, the unstudied intensity of uses at the Beach and its environs, conflict of uses, impeded use, conflict with the stated goals of the General Plan, noise, dog waste, aggressive dogs, health and safety, aesthetics, impacts on riparian habitat, impacts on native grasses, interference with migratory and resident birds, nesting and wildlife nursery sites, traffic, water quality, exposure to feces-borne bacteria, parasites and communicable disease, transfer of canine distemper to marine animals, and impact on Monterey Bay Marine Sanctuary.
>
> (Brandt-Hawley 2003: 6)

This list of problems is exhaustive and persuasive. It covers "health and safety" for humans and other animals (including the threats of bacteria and disease), dog movement (including "noise" and "waste"), and impact on birds and marine animals. Perhaps most interesting is the point called "aesthetics." Is it that canine companions disturb the environment and thus the aesthetic value, or are they just plain ugly (see below under "Dogginess")? Through these associations between dogs and their potential negative impact, they are produced as unnatural intruders in a protected habitat, and thus their allowability is called into question. Dogs and their liminal nature have no place in the "wild." The place image of Its Beach—the liminal space under scrutiny—is the naturally wild, in need of conservation practices. Interestingly, human cultural constructs such as traffic are included in the list of risks, not as a risk for the environment, but rather as something that the canine presence can interfere with.

The FOLF version of this issue is that although wildlife species should be protected—they are essentially something good—their degree of exposure is minimized. It is said that dogs mainly "chase after balls" and play with one another, and that there are only rare instances of dogs chasing after or killing other animals. Thus, the "wild" and the dog/human nexus may co-exist in harmony on the beach, and rarely interfere with one another. In fact, the canine presence may be beneficial to conservation, because it counteracts the presence of cats, which—in contrast to dogs—are said to threaten wildlife, especially birds:

> Birders generally acknowledge that cats in an urban area are a significant cause of bird mortality. We could find no information estimating the numbers of cats in and near the Field, but the presence of dogs in the park may actually act to somewhat discourage cat activity.
>
> (FOLF 2010)

This is an attempt not only to counter the accusations, but to widen the argument to include species not hitherto mentioned in the debate. Thus, dogs are presented as a solution to a problem not yet articulated.

Through this exchange of arguments, one can detect how the ecological matter of concern is being negotiated and contested, through the assembling and hierarchization of various actors and species. In the first version, a conflict over space is constructed in which dogs should not be allowed—their presence being a threat to other, more original species, which are worthy of protection. Version two produces dogs as part of the solution, rather than the problem. Interestingly, the second version does not enact the place as one of nature, where dogs could fall on the same side of the nature/culture boundary as wild species do. Instead, dogs are produced as culture and, as such, as belonging to particular parts of the beach: the seaside shore rather than the vegetation and the cliffs. Thus, the idea of pristine "natural resources" that are in need of protection is left untouched. One could instead imagine a situation in which this consecration is contested. What would have happened, for example, if the dog liberals had claimed that dogs should be part of this resource? Or humans? Or called into question the taken-for-granted idea that some species or habitats are in need of protection? Instead, a consensus take on "wilderness" as such is presented, the point of conflict instead being whether dogs contribute to or interfere with this wilderness. Conflicting notions aside, one issue that seems to connect both parties is that concerning the undesirability of dog "waste."

Excrements

If one spends any time watching people walking with their canine companions, one will note the colorful plastic bags that are carried along, attached to the leash, in pockets, or in specially designed bags. Dog bags come in various sizes and colors, in traditional plastic or biodegradable versions. "Dog poop" is perceived as a big problem in any urban environment, often debated in local newspapers and managed through, for example, containers designated for disposal. It is also a topic that both sides of this particular controversy engage with in what would appear to be great detail. Here in a blog posted at the RSCB home page:

> Because dog owners flagrantly ignore the leash and "clean up" ordinances, the place is practically unbearable for bodysurfers and walkers. Maybe this will wake up the dog owners to keep better control of their dog.
>
> (Rescue Santa Cruz Beaches 2004)

Dog owners are produced as dismissing regulations and not caring about the consequences, thus the appeal to prohibit off-leash recreation is their own fault, they should have seen it coming, since the place is "unbearable" to human visitors.

FOLF, on the other hand, stresses that the problem is minor, because the compliance rate, in comparison with usage frequency, is high (see also the quote from the animal police above). The organization is said to engage in regular clean-up days, when all members of the public are welcome. It is stated that, on such days, they also clean up trash from other visitors and tourists. "This includes both trash and dog feces, including what is left behind by our many out-of-town visitors, who are often unfamiliar with the regulations and expectations of the Santa Cruz community" (FOLF 2010). Through this statement, the question is widened to assemble not only dog dirt, but all kinds of waste, meaning that dogs are not the whole problem. All in all, Friends of the Lighthouse Field presents itself as a law abiding, community servicing, environmentally aware and responsible community. However, this is not the full picture, as will become clear later.

It is important to note the historical and cultural context of the dog dirt phenomenon. Susan McHugh writes that, following a report in 1979 on free-ranging dogs, there was a general turn in city policies across the US (2004: 178). The report highlighted the many health hazards these dogs presented in terms of fleas and other parasites, germs, and feces. Fear of contamination led to a new trend of leash and pick-up regulations (McHugh 2004: 178). However, the fear of dirt and feces is generally an urban one, with roots in the civilization process (Sennett 1994: 262). Matthew Gandy (2006a, 2006b) writes about the "bacteriological city," and discusses how public health has become the new institutional form of population control, leading to—following Foucault—hygiene and cleanliness becoming the new moral discourse and practice, as a means of bio-politics. Moreover, as Haraway notes, it is the human end of the trans-species couple who takes care of the dirt and thus is the prime object of this bio-politics (2003: 16).

Dogginess

It is not just the consequences and traces of dogs that provoke involvement, whether one is on the "for" or "against" side of the debate. There seems to be something about the nature of dogs in relation to the uniqueness of the space that triggers human emotions. As a telling example, in a petition to the Superior Court of the State of California, Santa Cruz, the LFBR writes this description of the nuisances that dogs on the beach (and in part their owners) present:

> Dog behaviors explained as incompatible with the recreational use of the Beach included constant barking, yowling, fighting, hunting for prey, chasing, digging, sniffing, begging for food, stealing food, running over people, shaking off water next to people, chasing and disturbing wild animals, marking of territory on towels and sand castles, defecating on the beach, owners kicking a thin layer of sand over the feces that lead to people stepping, sitting and playing in fouled areas, facing off against children who are playing at eye level, intimidation of children and other beach goers, and chasing skim boarders.
>
> (Lighthouse Field Beach Rescue 2005: 7)

The complaints include that dogs do not behave properly, as they do not behave like humans. They sniff and shake off water, they bark and fight. Here, it seems to

Figure 2.6 Dog on beach, Barry Island, UK (Photo: Joao Bento 2013)

be the sandy part of the beach that is worth protecting, in contrast to the argument above where the more "nature" side was produced as in need of protection. When the central actor is presented as human (people, children, skim boarders), it is the cultural beach that is of concern. One blogger notes that, "I also don't appreciate their crotch sniffing, barking, and abandonned [sic] excrement" (Rescue Santa Cruz Beaches 2004). In fact, it seems to be their perceived "dogginess" that is mostly disturbing, evoking feelings of repulsion. However, it may just as well be a matter of a lack of humanness, of civilization—humans generally do not defecate in public, they do not sniff one another's crotches or intimidate children. Thus, when producing the beach as cultural, allowability requires a certain amount or degree of civilization. The perceived lack of civilization also appears to be a matter of number, as one witness cited states: "on a typical sunny day during the summer it has not been uncommon to see as many as 60 dogs running loose on the beach throughout the day" (Lighthouse Field Beach Rescue 2005: 7). I will return to the number issue again and again in this book, as it is a central quality of humanimal crowding and trans-species crowds. For now, let me just point out that numbers create a certain sense of being out of control, "running loose," and moving forward in a way that is not rational. Again, "nature" is perceived in opposition to culture, and the larger the numbers of beings that are out of order, the less civilized the place. Numerous humans/dogs pose a threat to the ordered culture of the beach.

If the above statements frame and produce dogginess as negative, other personal stories accessed through blogs tell a different story. They are, among other things, about dogs' joy and motion:

> I swear dogs do smile. Famous Amos has a smile on his face right now just thinking about it. The people are really cool and great with their dogs. I've seen over 25 dogs on the beach without a single incident. The dogs are so happy.
>
> (Yelp 2006)

> By far and away my favorite dog park to take Bella to! Want to go to the beach, or to the dog park? Hey, why not both?! As much as Bella isn't a water dog, she loves the sand, so letting her run around while sitting on a blanket is a fabulous way to kill an afternoon.
>
> (Yelp 2008)

Here, the sandy part of the beach is clearly produced as canine. This is mainly accomplished by evoking positive feelings: smiling, happiness, the freedom to move. Control of animal movements is clearly also about letting them move (Lulka 2010); this is a counter-politics and a strong component to take into account if one wishes to understand why the unofficial "dog beach" can be maintained, despite the prohibition against unleashed dogs.

An ongoing controversy

The Its Beach controversy is not just ongoing at the discursive and official level, but the stories also contain information about local face-to-face incidents. The FOLF website, which is a very rich one, encourages people to actively work for a revised plan for the Lighthouse Field area, stating that, "it's time to balance the calls from the disgruntled anti-dog fanatics with a larger volume of calls from concerned citizens" (FOLF 2010). This could be done by calling a senior State Parks official, by participating in "cleaning the beach days," donating to the organization, and by acting as if the ban did not exist. "FOLF encourages you to continue to use Lighthouse Field State Beach (LFSB) for off-leash recreation during the traditional compromise hours (before 10 am and after 4 pm) regardless of the State's misguided new policy" (FOLF 2010). This time/space civil disobedience appeal is paired with a strong focus on the importance of being good citizens, of behaving properly, and acting politely when encountering officers and officials, to pick up dirt and to instruct others to do likewise. Moreover, its website displays hands-on suggestions for how to behave when confronted with a state park official. In short, the "politics of place" strategy proposed is to behave in a civilized manner at all times (except, perhaps, when playing together with dogs, a matter I return to below).

Interestingly, Its Beach is still understood by many visitors—connected or not connected to FOLF—as a dog beach, and talked about as one (filled with happy memories and harmonious encounters):

One day I speak to Anthony, a young man from Santa Cruz who regularly visits the beach along with his three year old "pit-bull-like" Sally, and he tells me how he got a ticket when Sally was only two months of age. He fought the ticket, and in the end, he got a reduced one, but the original fine was, according to Anthony, 300 USD. But he still sees this as a dog beach, and the ticket has not prevented him and Sally from visiting on a regular basis. I ask him why, and he says it is because Sally thinks it is fun, and when watching her play with five or six other dogs on the beach and into the waves, one can easily see why.

(Field notes April 14, 2010)

Today the weather is lovely, and an amazing amount of dogs in all shapes and sizes play around adults and children. There are no fights or conflicts to be found. I talked to a couple, Dave and Dianne, and they had no clue that dogs were not allowed off leash. This is also true for many other people I have spoken to, most people seem to think of this as a space for dogs, and often call it "the dog beach."

(Field notes April 16, 2010)

There is something left to say about the technologies of spatial control, and of social disobedience. First of all, the prohibition sign, visible at the entrance from West Cliff Drive, can be read as an expression of power, in which social control over citizens is asserted (Fiske 1989: 50). Interestingly, the text about dogs has been erased, so that it is impossible to discern from the sign what the rules are. Similarly, the dog leash can be understood as a technology of power and control over dogs. However, as sociologist Mike Michael points out, the leash is not only a device for handling the dog, it also controls the human's movement. The leash works as a technology to keep the cross-species bond intact (Michael 2000). In cases where there are no physical boundaries such as a fence, these technologies become more important (Instone and Mee 2011).

To sum up, this section has highlighted the emergence and continuation of a spatial conflict, leading to the creation of local interest organizations, involvement of city and state authorities and regulations, local news and social media, and, not least, park visitors (both humans and dogs, but also wildlife and cats). Through the debate, positions for and against dogs being off leash were consolidated, and I have pointed at several salient themes in the data: safety/risk, disturbance, excrements, and "dogginess," meaning the perceived nature of dogs. Moreover, the section has shown how the controversy seems to be an ongoing one, and that even though there is now a ban, there are actions being taken at various levels to try to restore the former time/space regulations.

Trans-species urban crowd

The people engaged in the interest organizations seem to be well equipped with capital, both economic and cultural. They know how to move around in the bureaucratic machinery and how to maximize the use of social media. The reason for the controversy described here cannot easily be placed "outside"—it is not social class, gender, race, or even dog ownership that can be used as an explanatory force. Instead, I have argued that understanding should be sought in the context of the debate. In the remaining discussion, I will consider the case in light of the concept of "politics of place," and develop the idea of local, collective action as more-than-human urban politics, by focusing in more detail on disorder and the movement of bodies, through the notion of the crowd.

First, I will expand on the notion of disorder, which is not understood as the opposite of order, but an integral constituent of it. The particular space, the beach in this case, is a liminal one, where disorder can be expected. However, there is a particular disorder that is not tolerated. It is not the children playing, or the surfers boarding, or even the birds noising that is sensed as disorderly, but in this story it is the dog/human nexus that creates concern; it is either perceived as wrongfully or insufficiently regulated. Furthermore, the disorder verbalized is overshadowed by the apparatus of ordering—the legal dispute, the prohibition signs, the hour regulations, the clean-up days—and the accounts from both sides bear witness to how "order and disorder are but flip sides of each other" (Weszkalnys 2007: 224). In the case under study, order/disorder is about cleanliness and matter out of place (dog dirt), but also about who belongs (people from the neighborhood, law-abiding citizens, dogs on/off leash) and who does not (drug users, homeless people, unleashed dogs, cats). In fact, the disgust that can be read in some accounts echoes the racialized discourse that is sometimes vocalized in talk about other urban strangers: suburban youngsters, punks, skateboarders, and homeless people (see e.g. Borden 2001; Weszkalnys 2007). However, "allowability" is produced by both sides through notions of humanness: civilized behavior, norm obedience, culture. This could be criticized from a sociological point of view. Conforming to public rules and behaving in accordance with them risk eliciting diversity. From an intersectional perspective, it is not only inclusion and exclusion colored by species that appear in the cityscape. But in addition, through the production of allowability based on qualities associated with certain humanness, class, race, and gender heterogeneity may be sociospatially reduced.

Richard Sennett notes with a similar argument that the emergence of modern city life—with fewer contact points and social arenas, along with suburban trends—has "eclipsed something of the essence of urban life—its diversity and possibilities for complex experience" (1970: 82). He argues that people need to feel some discomfort and "to experience a sense of dislocation in their lives" (1970: 160), and that disorder will ultimately lead to new structural changes. What is needed is not for the conflicts to be "solved" in any simple sense—following Sennett perhaps one can speak of anarchy in dogland. Because dogs are certainly different, the other in relation to humans, and still the same (by virtue of being human companions), their ambiguous presence as strangers is open to a number of interpretations and experiences—inclusions and exclusions that are not always comfortable or controllable.

Second, reading carefully the appeals and complaints, the matter of movement becomes a crucial factor. For the canine followers, dog feces and the well-being of ducks are seen as superficial excuses for restricting the animals from moving freely. Likewise, people who are afraid of dogs find themselves restricted in the presence of running and moving dogs. The movement of animals—animals in motion—as the dogs in this case, sometimes makes people feel disgust or causes fear, while for others it is a source of positive emotions such as joy or happiness. Action and movement are important dimensions in the urban production of meaning (Lefebvre 1996), and restriction of movement needs to be attended to by considering moving bodies and urban politics together. Sennett (1994) discusses the connection between various ideas of the body and the city in his historical account of Western civilization, where the fear of the horde and of uncontrollable collective behavior created a number of preventive technologies. I am tempted to interpret the dogs and their humans in this case study as a "trans-species urban crowd" that moves freely (well, within a few square acres) and is feared by some more order-inclined citizens. Thus, the number of arguments and the technologies used to control this trans-species crowd can be understood as following the Western tradition of managing collective bodies in the city, preventing riots through various forms of crowd control. Moving bodies in groups, including social movements, can be perceived as potential threats. Furthermore, dog/human crowds, similar to crowds of skateboarders, challenge the consumer-oriented culture, in that such crowds are neither consuming nor producing anything, but playing around only for the sake of pleasure (Borden 2001: 231). The benefits of adult people and dogs playing around in infantile ways—e.g. outside the rules and other framings of sports and similar activities—may be called into question. The beach is of course very much a site for play, and watching the action and movement of dogs and people at Its Beach is a reminder of the unruliness and bodily surplus energy that often comes out of more or less spontaneous action. Moreover, even though it can be argued that play is itself a product of consumer culture and capitalism, Rob Shields discusses whether the rational/libidinal binary may be a more apt frame than the labor/leisure one (1991: 94). In line with Bakhtin, the author draws on the notion of the carnivalesque in understanding the simultaneous resistance and compliance of moral codes at Brighton beach:

> The explosion of excessive behavior and social pleasures and leisure forms which is found in the seaside carnival is a mark of resistant bodies which at least temporarily escape or exceed moral propriety. Against the restraining empiricism, cerebral rationality, emphasis on control and economy, carnival produces a momentary social space based on the politics of pleasure and physical senses. [...] The grotesque carnival bodies on the beach are thus temporarily outside of social norms and embarked on a liminal project, even if they are in sites commercialised and territorialised in such a way as to control or contain any outbreaks of liminality.
>
> (Shields 1991: 94–95)

The trans-species crowd displays joy, play, happiness, running after balls—or described by the other side—as sniffing, barking, disobeying, running around in uncontrolled manners. Recalling the introduction to this chapter, in which I asked whether the humanimal crowd could be read as a threat towards a certain public order, this is where it gets tested. The carnivalesque nature of play at the beach certainly breaks with norms of interaction and thus constitutes a site for resistance. A particular feature of this urban crowd is that it also brings colorful plastic bags to the beach, the humans being polite to authorities and socializing with other visitors across the species barrier. Thus, the crowd portrayed in the present case study does not resemble the riots of the "badlands" that Mustafa Dikeç (2007) analyzes, in which poverty and violence are key ingredients, but rather constitutes a "goodland" riot, fought at the beach, in courtrooms, and in social and other media by actors well equipped with various forms of capital. Thus, Shield's approach to the carnivalesque does not apply perfectly to the case reviewed. The multi-species play performed on the beach does not suggest an escape from social norms or regulatory practice, but rather displays the simultaneous co-construction of rationality and order, pleasure and disorder.

Although this chapter can be read as yet another story of spatial conflict, it is also specific enough to not count as any history of urban practice (cf. Borden 2001: 265). It is the story of a particular suburban beach, where dogs inhabit a contested role as liminal creatures, roaming in a liminal place and, thus, forming a crowd together with their people, working both as potential threats to the social order, and as subversive actors who can potentially change norms and practices concerning our relations to other animals and the city. Thom van Dooren and Deborah Bird Rose ask whether "sharing" rather than "conflict" could be the plot for multi-species place-making:

> Places are materialized as historical and meaningful, and no place is produced by a singular vision of how it is or might be. In short, places are co-constituted in processes of overlapping and entangled "storying" in which different participants may have very different ideas about where we have come from and where we are going. [. . .] What would it mean to really share a place?
>
> (van Dooren and Rose 2012)

In Chapter 6, I will return to the promises of conviviality found in the work of various authors (van Dooren and Rose, Hinchliffe, Wolch). For now, I just want to point out some of the limitations of "sharing" as a metaphor, for example the risk one runs of obscuring power relations: differences in stakes and interests are problematic to acknowledge in human relations, yet even more so in multi-species ones (Holmberg 2011a: 209). Moreover, I am inclined to ask whether the prevailing consensus culture could be challenged. Contemporary politics, Chantal Mouffe argues, is too prone to finding consensus, so that actual arguing is closed down (1999). For her, a "good society" is not one in which differences are erased through technologies of consensus, but one in which thriving is all about allowing for heterogeneity and agonism. Following this line, would it be possible to call for a more-than-human politics of place, in which "dissensus" is the name of the game?

Summing up, "politics of place," as understood in urban studies, refers to struggles over meaning-making, practices, and definitions. Thinking about this process as a multi-species one, including intended and unintended actions and emotions, and as a process that involves assembling various human and non-human actors in and through spatial dimensions or qualities, enables a critique as well as a development of the concept of a more-than-human politics of place. Species-specific norms for how to behave in a certain place play crucial roles in producing allowability, through the process of exclusion and inclusion.

3 Stranger cats

Homelessness and ferality in the city

In the Disney film *Aristocats* (1970), bourgeois cat Duchess and her young ones represent common home values: familiarity, stability, belonging, and security. Stray cat Thomas O'Malley, on the other hand, is portrayed as free, unbounded, and witty. In the film, they meet up with the jazzy ScatCat band and break out jamming and dancing to the song "Everybody wants to be a cat." Cats, the most common pets, are in real life also defined in different ways based on both their relationship to humans (homeless, feral, domesticated) and the environment in which they move (domestic, indoor, farm, city). The cat is also surrounded, as with the Aristocats, by a lush symbolism of unruliness: cats are unfaithful, deceitful, sensual, and frivolous. In fables such as the story of the Master Cat— or "Puss in Boots"—he is portrayed as intelligent, savoir-faire and something of a trickster. Throughout history, cats have sometimes been associated with magic lives (Broberg 2004: 4). Feline symbolism is clearly gendered and sexualized, and the female cat is often depicted as mothering and caring (see Chapter 5). The cat is nowadays associated with the home, but this has not always been the case (Grier 2006). The Scandinavian cat was probably domesticated about 1,500 years ago (Broberg 2004), yet we know very little about its background. Gunnar Broberg (2004), who has written a history of the cat, argues that cats' paws step so lightly that they seldom leave any imprint on written history (Broberg 2004: 21). But we know that cats in pre-modern society lived in barns and on ships, and were primarily kept for their outstanding ability to catch and kill mice and rats. Not until the nineteenth and twentieth centuries did the cat, symbolically and materially, enter the home (Broberg 2004: 339). This shift also allowed for other, more intimate relationships with cats as pets (Grier 2006: 62). Although many cats today in fact spend most of their lives indoors, many of them live, at least part of the time, outside our control and scope of knowledge. Wandering through neighborhoods, they create their own feline spaces and relationships with people other than their own (Tipper 2010). In this chapter, I seek to understand how the spatial, social, and cultural orderings—the "cat-egorizations"— of cats play out in relation to home, street, and other places, as well as in relation to actors who handle urban cats and, not least, legislative frameworks. The home is, like the street, a place that is both symbolically loaded and politically charged (Blunt and Dowling 2006).

Cats' popularity is not always beneficial for the cats themselves, and in contrast to myths and beliefs about the filthy, treacherous, and magical pussy-cat, she is also genuinely vulnerable in society. Cats are common (according to a survey, there were 1.3 million cats housed in Sweden in 2006 (Statistiska Centralbyrån 2006)—a country with a human population of just above 9 million), but have low status. Each municipality has its own cat charity and shelter, caring for and relocating homeless cats. In Sweden, the law of supervision over dogs and cats regulates cat ownership and supervision issues (the Dog and Cat Supervision Act 2007: 1150 came in to force in 2008), and the Animal Welfare Act (1988) regulates the care of the animals—with minimal standards for what is regarded as good pet keeping. It is difficult to find statistics, but sources refer to 100,000 homeless cats in Sweden. Nevertheless, it is clear that, among the authorities I have interviewed, the "cat issue," especially in urban areas, is produced as a growing social problem. One way of handling the problem by reducing the number of strays is through a range of adoption practices, managed by public or private shelters. Julie Urbanik writes about the situation in the United States:

> Today many people know that adopting an animal from a shelter saves an animal's life, but many people do not realize the full extent of the problem of dog and cat overpopulation. The HSUS [*Humane Society of the United States*] estimates there are around 70 million stray cats and dogs roaming the country at any given time, due to abandonment, uncontrolled breeding, and running away/escaping. For animals who do make it to a shelter (between 6–8 million), only about half (3–4 million) are adopted out. The other 3–4 million are euthanized. That is anywhere from 342–456 dogs and cats per hour 365 days a year. It is not only a heavy emotional toll on the humans who have to "put down" such a constant stream of living beings that are otherwise healthy, but it is also an increasing financial burden to local state, county, city, and private shelters.
>
> (Urbanik 2009: 43)

This is clearly a bad situation, one in which both cats and caring humans suffer. Studies have shown how shelter workers and veterinarians learn to cope with the killing of unwanted pets (Patronek et al. 1996; Balcom and Arluke 2001). This chapter, as a contrast, looks further upstream at the process of making pets killable, zooming in on the management of cats that roam free on urban streets, which are liminal spaces indeed. Cats that do not belong to a home are anomalies in need of correction. This management process sometimes leads to the production of stray cats as a social problem that can be dealt with in a calculating way, thereby neglecting our human/feline response-ability (Haraway 2008a: 80).

Figure 3.1 Street cat, Lanzarote (Photo: Tora Holmberg 2014)

On the basis of news reports, popular cultural images, interviews with animal police, animal welfare inspectors, cat rescuers and shelter workers, as well as complaints and other documents, I investigate a number of controversial positions related to the process of "becoming with many" cats (Haraway 2008a: 4), and the norms of appropriate cat keeping in the city. The phenomenon of urban cat management raises a number of questions: Where and how are "homeless," "feral," and other "cat-egories" produced, and with what consequences? What is the role of "home" and other places in defining and handling various categories of humanimal relations? Regarding the terminology of zoocities and humanimal crowding—the overall tropes used in this book—there are several points that this chapter will highlight. First, the urban context provides certain logics, where meaning is ascribed different "cat-egories" in relation to the place in question, and where ordering a perceived disorder using different methods is at the core. Second, there are, besides the cat-egorization practices, numerous other technologies of crowd control involved in urban cat management. Before moving on to the case study, which is presented in the two sections "Homelessness and crowding" and "The feral and the free," I will provide some additional background on the topic, concerning both feline home/homelessness and cat spaces. In the concluding section, I return to the questions posed above, and discuss how anomalies such as homelessness and ferality are normalized, thus rescuing the image of the sacred home and the cat as a tame and cherished pet through a one-owner-one-cat model, while the wandering cat as an urban "stranger" may shake this modern image.

Home and homelessness

In order to make sense of the study, I use the probing concept of "home," in particular the lack of home. As stated by Christopher Jencks (1994), in order to understand homelessness as a phenomenon, we need to scrutinize the meaning of home. The home is historically the prime place to raise future citizens. It is packed with ideas of purity, shelter, exclusivity, and family values, but also feelings of belonging. According to Doreen Massey, the meaning of home is, as with other places, a result of culturally and politically charged negotiations of time and space: "That place called home was never an unmediated experience" (Massey 1994: 164). "Home" is produced through place-making and identity processes. Pets are often deeply intertwined in these processes, in their owners' notions of home, and act in negotiating the meanings associated with it, not least in terms of belonging (Fudge 2008). Pets are indeed often regarded as kin and members of the family (Mason and Tipper 2008), providing social support similar to the human members (Bonas et al. 2000). However, as pointed out by Rebekah Fox and Katie Walsh, their membership is circumscribed, and their status as belonging in the legal sense becomes clear when circumstances for the humans of the family change, for example through migration (2011: 114). The arrival of new children, moving into new neighborhoods, or illness may jeopardize the position of the non-human family member, who is sometimes left behind as a stray.

   The politics of human homelessness has been well covered from a social science perspective (Jencks 1994; Hutson and Clapman 1999; Kennett and Marsh 1999). Essentially, the literature points out that various explanations of homelessness—whether structural (lack of housing, poverty, welfare state cuts, etc.) or individual (drug addiction, unemployment, mental illness)—suggest their own solutions and interventions. It is often emphasized that homelessness is less a state than a process, where a "cycle of homelessness" can be identified (Wolch et al. 1988: 443). People move in and out of shelters, prisons, and other institutions, to friends, back to the shelter, and so on. Moreover, public shelters constitute "hybrid organisations" (Hopper 1990), unable to solve conflicting demands to officially serve as institutions for acute shelter, while in reality providing more or less permanent housing. From a critical point of view, it has been argued that the shelter system is a node in the apparatus of homelessness, co-constructing and allowing this social problem to continue.

> The homeless industry is outside as well as inside the homeless culture, reinforcing that culture while at the same time being dedicated to its destruction. It lifts people up only, in some cases, to put them down—or, to put it the other way round, it puts people down only to lift them up.
>
> (Somerville 2013: 403–404)

Thus, homelessness is a multi-dimensional, "storied," and socially produced process, rather than a given social problem. When it comes to feline homelessness, there is a growing volume of literature on the shelter movement, especially in the US. Its history of caring for strays, its connection to the humane societies of the late nineteenth century, and its role in formulating the social problem, is intriguing (Irvine 2003). The social problem has been framed as animal surplus—in the US around four million cats are killed annually by shelter workers alone (Patronek et al. 1996)—but where the blame for this situation is placed has varied over time. Currently, irresponsible owners are on the agenda, but many shelters hesitate to overtly blame owners (Irvine 2003). Thus, animal rescue and adoption arecertainly a "negotiated order" (Balcom and Arluke 2001). I borrow the term "hybrid organization" to characterize the cat shelter, an organization in which conflicting practices such as advice and courses on behavioral problems, veterinary care, neutering, euthanasia, and adoption, take place under the same roof. The national context of the study is Swedish, where there are countless shelters—every small town has its cat enthusiasts. Some shelters have contracts with the local police, according to which they house "adoptable" homeless cats for a fee, thus official institutions such as the police and animal welfare authorities are important nodes in running the hybrid organization. My question in this chapter is: What is the meaning of home—the place-image (Shields 1991)—in the context of cat homelessness and ferality? Moreover, I ask how the figure of humanimal crowding is involved in the understanding and management of homeless and feral cats.

Cat spaces: wild, feral, domestic

In the preceding chapter, I referred to Simmel and his notion of spatial ordering through boundary work and categorizations as a central dimension of urban sociation (Simmel et al. 1997: 138). As Lefebvre later pointed out, the social production of space can be analytically separated into three different dimensions, in dialectical relation to one another. Social and physical space are produced through various practices of everyday life, and emplaced in specific locations. Thus, the homely space, as imagined through symbolic structures, is emplaced by various actors. And, as Lefebvre points out, the social production of space is also a means of gaining knowledge, power and control (1991: 26), through the process of "spatialization" (Shields 1991). In my view, animal involvements in these spatializations are symbolic, social and material, and their ambiguous presence and action must be regulated in public, to avoid rendering a notion of disorder, of "insanity of place" (Lefebvre 1972: 389).

> The place of the animal in the city is uncertain and often contested, rather than being determined by epistemic principles. These principles constitute

the basis of a power structure or an ordering system, but one that is clearly resisted at the local level.

(Griffiths et al. 2000: 61)

The ordering of animals in and through urban space is further complicated by the notion of the "feral," typically addressing non-native populations of species. In that sense, they accommodate an out of place-ness, as anomalous, and "stand for wider aspects of social anxiety and danger" (Franklin 2014: 139). As such, the feral cat is a provocative anomalous other, and according to Griffiths et al. (2000), it can be reframed into two possible identities—either as wild or potentially domestic, or as tame. Domestication is a relation of kinship, involving taxonomy, intimacy, power/property, and control of reproduction (Russell 2007). Symbolically and historically, domestication is connected to household and home, and different categories of animals—wild, tame, game—are often placed differently depending on their proximity or distance to humans. But domestication is about relationships, "an emergent process of co-habiting, involving agencies of many sorts" (Haraway 2003: 30). Thus, neither humans nor other animals are easily placed and they do not stay put. They move about, transgress and disturb taxonomies and orderings. As suggested in the preceding chapters, human/animal relationalities and becoming-with-many are emplaced processes, taking place and being experienced through the dialectical relation between form and experience, space and subjectivity.

Homelessness and crowded streets

According to my interviews with police and animal welfare inspectors in Sweden, the "problem of unwanted pets" (Irvine 2003), in particular cats, is one of the top three most common urban sources of animal related controversy, along with animal hoarding and dangerous/aggressive/unleashed dogs. It is also something that consumes time, personnel and economic resources, and needless to say, involves considerable suffering. Listening to the stories provided by interviewees, the problem is in essence that cats, compared to other pets, inhabit a rather strange place in society, as explicated in this quote from a police officer:

Interviewee: So the eh, cat problem is rather big and it . . . I don't know, I think it's a matter of attitudes. A cat, you let them out like this [

Tora:] uh-huh

I: people just think that the cat is supposed to run free, and, eh . . . if I have to decide, you'd only be allowed to have cats indoors . . . in the urban environment [

T:] in the city?

I: Yeah, absolutely. I mean, it's not stranger than if you have a guinea pig or a hamster, you don't let it out in the morning and collect it in the evening. But . . . and the dog, you don't let it out, but you take it out for a walk [

T:] uh

I: And if you think that the cat needs to get out, then you'll just have to walk with it, right. So here we have a big problem, actually, and . . . we haven't come up with a really good way of making everyone understand this [

T:] uh

I: because it's enough that some people are careless [laugh] then soon there will be many cats, 'cause the cat shelters are overcrowded.

(Interview police officer)

The police officer in this quote tries to make sense of the cat overpopulation problem, and his explanation is that people let their cats out to wander around, and implicitly, they also neglect to neuter their animals. If we would treat cats like dogs, for example, there would simply not be any problems—at least not with cats crowding the streets of the city. There is however a complicating issue here: the cats who run around do not always belong to an owner. Thus, the police need to distinguish between different species of cats, which form three different "categories": the lost, the homeless, and the feral. These categories are interesting because they are produced through an assemblage of different jurisdictions, formal definitions, and interventions. However, as will become clear, this ordering is not easily done. The first category is the lost cat, salient when discussing a range of urban-related animal controversies handled by the police:

I: Yes, and then lost animals, of course. You [

T:] Right, that people bring in or . . .

I: Yes, yes, often that people bring in, or that they, they call and say that "we . . . this cat, it's come and lived here now for three weeks in our carport and we have put up notes at ICA [local grocery store] and everywhere and no one has called, so we . . . this one seems abandoned."

(Interview police officer)

Cats, dogs, rabbits, and snakes are sometimes lost, or can be assumed to be lost, and they are taken in to the police station or straight to a shelter. Pets that are earmarked or equipped with a microchip can easily be identified and, often, returned to their owners quite quickly. This category is regulated through the Lost Property Act, and handled exclusively by the police. Other actors are of course the cat and the (absent) owner. Note how the cat in the quote above is assumed to be lost somewhere close to where it was found, like a wallet or a key. The movements of cats are seldom acknowledged, although we sometimes read stories of cats who get lost in one location and are found years later, somewhere completely out of place. The *Guardian* for example, wrote the following about Willow:

Missing Colorado cat found in New York

   A cat that went missing in Colorado five years ago was found wandering in Manhattan, and will soon be sent on a plane to reunite it with its former owners, an animal pound spokesman said. Workers at the pet shelter traced Willow the cat back to a family in Colorado, thanks to a microchip embedded in the animal's neck that they checked with a scanner, said Richard Gentles, spokesman for Animal Care & Control of New York City.

(*Guardian* 2011)

The second cat-egory is the homeless cat or, in rare cases, the dog (dogs are seldom overtly homeless in Sweden, due in part to almost a century of dog tax, from 1996 exchanged with registration and marking regulations). If the lost cat is not returned to an owner, it is categorized as homeless.

I: The homeless cats, on the other hand, they can have run away or they can have been dumped. And they . . . we try to, to eh, pass them on to a new home.

(Interview police officer)

Homeless cats are said to be adoptable, or it is at least the aim of interventions made to make them such. Thus, homeless cats are not shot on the spot, but examined by a veterinarian and placed in a shelter for future adoption (more on adoption practices below). This category involves the cat, the police, the shelter, and the future owner, and is regulated by the Dog and Cat Supervision Act (2007: 1150).

   The categories "lost" and "homeless" are thus produced through how they fit in to different jurisdictions, and it can be financially advantageous and easier on the workload to place the cat in the "right" category. I asked how the police officer thinks the system works today:

I: Well, I think eh . . . if we talk about cats for a while, it's a curse, it almost knocks you out, right. There are wild cats and abandoned cats and lost property all the time. [. . .] They bring in cats and, eh, they call us and say that they have taken in a cat, and so they book them as lost property . . . and then we are responsible for caring for them, for ten days as it's written [

T:] uh

I: That's 500 bucks [spänn = SEK] a cat. And bills of 20–30,000 arrive now and then. [. . .] Regarding the cats, there are three different jurisdictions to turn to, right. The Animal Welfare Act, the Lost Property Act and then the Supervision Act. We have to toughen up, we can't book everything as lost property. Is it marked, the cat, then you would probably consider it as a lost property [

T:] uh

I: is there an owner then, then it can be the Animal Welfare Act. But if they are not marked and they are a little skinny and a little rough and ear mites and the like then, then you can like [sigh] suppose that they are abandoned.

T: Uh.

I: Formerly it was written that they, eh, should be, could be suspected to be abandoned or feral. Now it says . . . no, *and* it was then, now it's *or* . . . So it makes it a lot easier and can be "suspected" to be [feral], that's like very low on the scale of suspicion.

(Interview police officer)

Details in the writing of the law give suggestions as to how to define the cat according to the categories available, suggestions that the police officer in this instance needs to be able to translate and interpret in order to make things work smoothly—in short, to make the procedure accountable. The homeless cat is defined as one born in a home with an owner, but who has been abandoned or run away, thus becoming an object of suspicion and corrective efforts. And, as the quote above shows, the boundary of "homeless" and "feral" is a slippery one, yet equally important because the responsible actors as well as the future fate of the cat vary with the category:

I: Mm. . . . Ehm, we try to find some way of drawing the line . . . between these two concepts. What is a homeless cat and, and what is a feral cat, because feral cats you are not allowed to enclose, right. This has been stated by the Ministry of Agriculture [

T:] hm

I: that it, it's quite demanding, psychologically, on the animals. Ehm . . . so when we trap them, there are these cat populations that are feral, right, clearly [

T:] they are like born in the wild?

I: Yes, they are born outdoors.

T: Mm.

(Interview police officer)

Note how I, as the interviewer, and the interviewee both try to establish whether the cat has been born in the wild, thus producing origin as a key to cat-egorization as feral. However, this origin must be deduced from a number of signs, and the categorization of the feral cat is typically based on a number of visual cues. Here I ask one of the shelter workers, how on earth you can distinguish a feral from a homeless being:

I: It's sometimes a little hard to see because you cannot get close to that cat. And it's still the most reliable, or it's not so reliable either, but if you look at a cat that's lived outdoors for a long time, it often looks rather rough. You can see it by the walk, it's often quite sway-back too. Because they, they walk low.

T: Mm.

I: Because they, they are used to . . . they try to make themselves as small as possible. That's a rather reliable sign [

T:] mm

I: I'd say.

(Interview shelter worker)

The feral cat looks a bit rough, it walks low and has a sway-back, but there may be other cues available as well:

T: I was thinking of this too, with cats and homeless or feral cats. That is also a problem, right, I [

I:]that's a huge problem

T: assume it's a problem.

I: A huge problem. What is a feral cat?

T: Yes. Where do you draw the line?

I: Where do you draw the line? It can be fearful.

T: Mm.

I: But it can also be feral.

T: Mm.

I: But we say like eh . . . people shouldn't feed cats, but if you see, they call here and ask. If you see eh, a cat that brushes past the house then, you call the city hunter and then they put out a trap and then they get to check the cat. Is it marked, does it have a collar or something, chip? And is it healthy or not, right? The, then they simply take it away. It is, eh . . . they can see if it is feral, if they notice that it's a house cat, they'll leave it to one of our cat- or dog shelters. And then we'll see if it's searched for. [. . .]

T: Mm.

I: What can we do, then? We can't fill our cat and dog shelters with, with, eh, cats, it's not possible. But, then, either Pelle is rehomed, if he's not marked or chipped he is rehomed or put to death, it's as harsh as that.

(Interview police officer)

Here, the fictive cat named Pelle is used to highlight the process: his fate is either to be redeployed or killed. Based on a number of markers, whether he is chipped and looks healthy or not, or behaves fearfully, he can be understood as either homeless or feral. Interestingly, the fact that shelters are crowded with cats is used as an argument to kill the feral ones. But there are also other arguments used. The police officer tells me what happens when someone calls in to report an abandoned cat:

I: Most often it's common house cats, not any valuable cats. And, of course you make some sort of investigation first, like, ask them to check how long the cat has been around, advertise in the neighborhood and . . . see if there is any reply. And if not, one can take away, it's important that it is the right individual. So, you need to make sure that the one who takes away the cat, contacts the complainer so that it is the right cat.

T: But, to take away, then you mean that they, they kill them?

I: Yes, that's what you do.

T: Straight away? Uh.

I: That's, right [sigh], and that's also something that we, we at least would want them to do.

(Interview police officer)

Put bluntly, cats are cheap, they often totally lack any economic value. When negotiating the place and category of a particular individual that seems to be homeless or feral, this fact becomes an important "indexical" factor (see below). And, when it comes to handling abandoned cats, the costs may be substantial for the authorities. If the authorities choose to place them in a cat shelter, at least some costs need to be covered by, in this case, the police. Thus the police officer in this quote hopes that the cat can be "taken away" at this first instance. As will become clear in the next section, some interviewees similarly argue in favor of euthanasia, adding that it is not in the best interest of the feral cat to be housed with humans—because they are not tame, they dislike human contact. While stray dogs are always—if not aggressive and attacking authorities—placed at a shelter, city wildlife managers may in the end be authorized to shoot off or in other ways kill abandoned and feral cats.

Summing up, the categorizations of anomalies, with consequences with regard to the cats' future fate, point at a construction of the "normal" case. Normality is produced here as a contrast to the anomalies of the lost, the homeless and the feral, e.g. the house cat, the cat living in a home with one responsible owner. The handling of these anomalies aims at normalizing the established order, either through returning the cats to their owners, killing them or sheltering them for future adoption. It may seem like a straightforward operation, but the above analyses of the classification practices suggest what in ethno-methodological terms is called "indexicality" (Lynch 1993: 19). This simply means that there is no way of defining once and for all the true or objective meaning of a word or a phenomenon, its meaning is always negotiated and defined by its context, or rather by the context that people deploy in order to create the meaning of the word (Coulon 1995: 16). However, members transform, through a number of measurements, the indexical categories into objective ones (Kumlin 2011). In this case, lost, homeless, and feral are defined and made meaningful in the context of the legal regulations, the actors involved and, not least, the place of the action. Consequently, "urban," "street," "shelter," as well as "home," "lap," and "garden," play important roles in constructing meaning around the cat by emplacing him/her.

The feral and the free?

Thus far, I have argued that the urban cat is "cat-egorized" into different juridical spaces in accordance with its appearance: its feline belongings are marked by bodily gestures in relation to the place it inhabits. Moreover, categorizations are not innocent practices, instead knowledge/power and value/practice are structuring pairs that go hand in hand. The feral cat is viewed as implicitly bad, as an anomaly that is in need of either euthanasia or transformation through taming. However, as stated in the introduction to this chapter, cat symbolism is also full of positive connotations surrounding the untamed: as free, wandering, and frivolous. The question, then, is whether there is such a being as a fully tame cat. Some of my informants also rehearse the idea that, "you don't own a cat, you are cat owned." In the following, I will investigate the paradox of the tamed feral cat through two forms of urban crowd technologies—control via the neutering/spaying of cat colonies and taming via domestication. Relating to the discussion in the previous chapter concerning use of the leash, these technologies can be understood as related to different power regimes and ideologies. While neutering/spaying is an example of disciplinary power, intervening invasively with the body and its reproductive capacity, taming through domestication practices is a technology that connotes a governmentality-oriented regime, in which the will to act in a certain, preferred way is facilitated (see Foucault 1991).

Feral creatures are interesting. They disrupt—and have perhaps always disturbed—cultural sorting systems and are interpreted as out of place. However, ferality is not a given quality; its out-of-place-ness is produced by various actors in relation to current norms and politically charged matters. In *Savage Girls and Wild Boys* (2002), Michael Newton writes that one common feature in the history of feral children is that they rock notions of what it means to be human. Myths about children who have been raised by wolves or dogs, or in captivity deprived of any social contact, are deployed to strengthen different contemporary ideologies of humanness. Thus, when the idea of the noble savage was in fashion, narratives were thought to confirm the purity of these children, untouched by human society. Because they are neither human nor animal, or perhaps both, they are validated as proof of both the nature and the nurture poles of the humanness debate. So what are they, exactly? In Newton's words, they are not outsiders—beings who participate, but from the margin—instead they are liminal figures, as in the following quote analyzing Jonathan Swift's narration of a feral boy: "Peter, on the other hand, belonged nowhere, and looked on all human cultures from an edge, a liminal place, where even the fact of his being human at all stood in doubt" (Newton 2002: 40).

In a similar manner, homeless people living under less-than-human conditions, like the "mole people" of the tunnels in New York City, are sometimes conceptualized as feral, wild, and untamed (Toth 1993: 1–6). Like the pet, a feral human is neither human nor animal, neither culture nor nature. Instead, the feral creature—human as well as feline—occupies a liminal space, becoming both and in between. Let us zoom in to this space.

Feral cats—like their human counterparts—often occupy more or less abandoned places close to human dwellings: closed factories, backyards, alleys. They roam garbage bins and hunt prey in order to survive. Moreover, people living nearby or passing the area on a regular basis often stop and feed the cats. Cat shelter workers engage in their wellbeing, feeding and neutering them and, when necessary, making sure they receive veterinary care. However, as noted in cities as diverse as London and Singapore, feral feeders endure a great deal of trouble and risk to care for these animals (Griffiths et al. 2000; Davies 2011). Needless to say, owing to their lack of protection and care, feral cats suffer from stress, health problems such as mites and parasites, starvation, constant pregnancies and births, and threats from predators, especially humans. Neighbors often find the cats disturbing, conceptualizing them as vermin due to their noise and numbers, but sometimes also empathize with them and report their suffering to the authorities. Thus cat colonies present themselves to numerous actors. So, how are they handled?

One technology for handling feral cats is taming. Taming and domestication are relational processes. In the autobiographical novel, *A Street Cat Named Bob* (2012), the author James Bowen, himself with a history of homelessness, tells the story of how he became a different person after meeting a red stray cat and being escorted by him through the streets of London. James supports himself by playing the guitar in public spaces: outside metro stations, on squares and sidewalks populated by tourists. Bob chooses James's company and home, while James finds himself being tamed: letting Bob out at certain intervals, giving up territory and choosing sites to play that are Bob-friendly. Using Vinciane Despret's terminology, human/animal interrelations such as Bob and James's, are made up of "anthropo-zoo-genetic" practices (2004), through which bodies of different kinds become more available to one another, more attuned to one another, and thus reciprocally different (see also Weaver 2013 on "becoming in kind").

Whether feral cats can be tamed is a matter of controversy between the authorities and shelter workers interviewed, as well as between different shelters. The point of conflict is, in short, the question of whether or not feral cats are tamable. From the perspective of animal welfare inspectors, whose utmost responsibility is to make sure that the animal welfare regulation is followed, the answer is "no," because the taming of ferals includes a certain amount of suffering.

I: I'm of the opinion, and this is my personal opinion, that feral cats should not be put in a cage. Sure, it's possible to tame them, but, it takes such a long time. . . . My mother has lived on a farm and, it, cats find their way to animal lovers [

T:] uh

I: and it's, she put out a bowl for feral cats in the neighborhood and thought that "poor kitty cat, well, you shouldn't be afraid of me, but I put out food tonight and then I turn on the heat cushion in here and you can crawl into the shed." And after a year she gets to pet them [

T:] uh

I: and so she knows they've had food, food and water and heat . . . but she hasn't wanted me to take them away but . . . eh, it, ok then . . . those who've chosen to be with her, they've been allowed to be there and have had a good home, so sure, it's possible.

(Interview animal welfare inspector)

Taming feral cats is, according to the above quote, a time-consuming attempt at forced domestication, the success of which depends on a large amount of patience on the part of the human ("it takes a long time") and on the agency of the cat ("those who've chosen to be with her"). In short, this process seems to involve in large part earning trust through offering food, warmth and care. As a reward, perhaps after a year or so, you may be trusted enough to pet the cat. The quote aptly illustrates the idea behind the concept of "domestication," as a relational, historical, and situated process constituted through practices and invested with work, attachments, and emotions. What is at stake in the relationship is the "becoming with" other species (Haraway 2008a). But, make no mistake, this becoming with is not an idealized process. On the contrary, it is, using Haraway's terms, ongoing within "the complexities of instrumental relations and structures of power," in which actors are both the means and ends to one another (2008a: 207). However, this time-consuming process is not what most shelter workers have in mind, when the organizational—financial and spatial—limitations of the shelter are the reality they need to adapt to.

I: 'Course some work with saving all cats.
T: Uh.
I: Or all animals. We don't have, we cannot do that. We take care of these cats that we believe will get a home. That have a chance [
T:] yeah, right, yeah . . . yeah.

(Interview cat shelter worker)

Given that there is a limited amount of space at the shelter, it is better, according to this interviewee, to bet on the cats that have a chance—that is, the tamer ones—and let the others be killed. This view is contested by colleagues, claiming that it is not fair to blame the cat for its bad manners:

I: So but, we would like to [sigh] give all cats a chance. Just because you've had a shitty life to begin with, right [
T:] yeah
I: it shouldn't mean that you don't get a chance, everyone gets a chance. 'Cause we've had cats that, the first year they've been . . . well, quite feral, you know, they've spat, you, you haven't even been able to look at them. And then suddenly one day, it's turned around and they've become a lap cat and today they're the cuddliest cats. When they finally realize that we're not dangerous, then . . . they can become, those are the finest cats, I think.

(Interview cat shelter worker)

In both of the quotes above, the bottom line seems to concern the cat's "homeability"—as a continuum of homeliness—however, the second shelter worker talks about the process of taming as a possible option, eventually resulting in a "lap cat." When asked for some good examples, the same shelter worker mentions a particular feral cat who was successfully rehomed.

I: Eh, and it's so damn nice because we still get emails from her [the new owner] and the like, where she kind of tells us how it's going, and the cat, the cat sleeps in her bed, the cat plays with the poodle and the cat is out walking in the garden and so on [

T:] [laugh]

I: It truly works very well. It's like, even those [feral] cats can actually, in the right environment become . . . quite good cats.

(Interview cat shelter worker)

The feral cat is lost in a moral sense but, with a little bit of matchmaking, even the most troublesome character might—according to the interviewee—eventually become a "good cat." The quote thus exemplifies a moral discourse on cats, in which the tame cat is good, with the feral creature thus being implicitly bad. As a contrast, the successfully transformed feral cat in this example sleeps in a bed, plays with other members of the family, and occasionally strolls in the garden. She is completely tamed.

However, to complicate the picture even more, there is also a strong discourse around what is morally wrong about taming a free creature. If the cat chooses to live at a distance, and thus dislikes any closer human contact, why should they be tamed with force? The following is from an interview with another shelter worker:

I: Yes, to try to tame them. And that, I think it's such a distressing process that it's, well . . . I'm not saying that it's better to kill them, 'cause it's like, to some it can, they can absolutely have a good life and become tame or accept living with humans but . . . my God, it's like . . . eh . . .

T: Yeah, there are also many different . . . opinions, right [

I:] right

T: around this issue. If it's possible or not to [

I:] to tame wild cats?

T: Mm.

I: Yeah. Yes, I think it's . . . if it can be avoided, then let them have their free life if it's possible, right. Let them live wildly, it's not the dream of all cats to be tame and it's not the dream of all cats to lie on a sofa and purr, you know.

T: Mm.

I: Just because they're cats. They can actually . . . have a really good life . . . even when they're feral.

(Interview cat shelter worker)

Through this discourse of the free and the feral, cats are articulated as actors who may choose to live in the wild. But the dilemma remains: feral cats suffer in ways that go against both the animal welfare law and people's moral judgments. One alternative used to control feral cat colonies by cat advocates like the one above is the trap–neuter–release (TNR) method. It is used to control the reproduction of the population (so that the colonies do not become overcrowded), while letting the cats maintain their existence at a certain distance from humans. It has come to be practiced worldwide, and the *South China Morning Post* printed a witty report on the cat colony program in Hong Kong:

> Snip for strays helps win feral cat fight
> The Society for the Prevention of Cruelty to Animals (SPCA) set up its Cat Colony Care Programme in 2000. The method behind its attempts to stabilise and reduce the number of stray and feral cats involves three steps: trap, neuter and release. The rise in the number of homeless cats it has treated has been dramatic, and the results have been impressive. In 2000, 355 cats were handled. The figure last year was 6,317. "There's a decreasing trend concerning complaints about these cats. We're having to euthanise fewer animals and we're seeing fewer sick cats," said Fiona Woodhouse, the SPCA's deputy director of welfare. "It also shows that it has been effective in drawing public participation to control the stray cat population."
> (*South China Morning Post* 2013)

The neutering and spaying of wild cat populations can be viewed as a bio-political tool, used to ensure that the colony does not multiply. The trap–neuter–release program, as spelled out above, has had some impact in Sweden as well. It is advocated by some cat shelters and charities, but not by others:

T: But there is no common . . . view of this [TNR] here in Sweden?
I: No, no, and some cat shelters have been doing this for a long time kind of and [
T:] uh
I: their county administrative boards . . . know about it and don't say anything, or, like, don't think it's a problem.
(Interview cat shelter worker)

In some Swedish cities, feral colonies are cared for unofficially by cat feeders and volunteers working for charities, including veterinarians who perform neutering almost free of charge. However, officials do not see this as a preferable practice. When asked what to do about feral cats, one of the interviewees, head of the county administrative board's animal welfare section, states:

I: It's really difficult, I don't really like this trap–neuter–return method, eh, because it's, of, it's really detailed that the animals should have the daily inspection and so on, they are not getting that really [
T:] uh
I: I think. That's my opinion. If you release, sure, they are not getting more. That's good, right. But, then when they are ill, who will notice that? That's how I think.
T: Mm. There still needs to be someone who is responsible for them [
I:] right, true

T: and gives them supervision.

I: In a sense they get that when someone feeds them but, it's not like, it's not on an individual basis [

T:] uh

I: still, if it's about a population of 50 cats, you can't, it's not certain that they'll all be there, right, when you're there to feed them.

(Interview animal welfare inspector)

The interviewee clearly objects to the TNR method, because it does not itself solve the problems concerning the cats in question. In addition, it poses a problem for the practitioners: they cannot see to and monitor each and every cat on a regular basis. The collective, in this case the cat colony, is an anomaly according to the law, which the inspectors are there to enforce. Here, a similar objection from a different inspector, also head of a county administrative board animal welfare section, is clear as regards the legal complications:

T: In certain places they've tried to, right, to bring in the cats and spay them, neuter them and then release them again.

I: Well, then I think you have taken on an owner responsibility, because here you also have the post-surgery care, first of all, for the neutered animals and I've been asked to be part of these . . . trap, neutering [

T:] trap, mm

I: release, TNR projects, and I've felt that eh . . . it's not for me, 'cause then, I take on, the responsibility of . . . costs and labor and to . . . in order to do these things then, should, you must adhere to the animal welfare act in what happens after too and, then you can't just release the cats.

(Interview animal welfare inspector)

In Sweden, the Animal Welfare Act (1988) clearly states that all domestic animals should be cared for and supervised on a daily basis, by a defined owner/holder—reindeer being a telling exception, as they are allowed to be free ranging during the summer months and graze in the mountain areas of northwestern Sweden. For the welfare inspector above, a technology such as TNR is unthinkable because it breaks the law, as there is no one to hold responsible for feral cats, no matter how many feeders and dedicated vets there may be to care for the colony. Caring for a collective is not a solution, because the law refers to individuals. Crowds are thus not conceivable in legal terms. To put it bluntly, animals such as cats are required by law to be owned, and if they are not, they are not allowed to exist. They are non-citizens and thus made killable, as if socially prepared for death (Haraway 2008a: 80). The funny thing is that, as the animal welfare officer states above, if and when you start feeding a cat, you may be held responsible for it, through a kind of acquired or earned ownership. This is also brought out in the quote below from an interview with a police officer active in a city of medium size, according to Swedish standards. In addition, he describes how cats themselves are active partners in this process:

I: If they in fact feed a cat, then it's written in the white papers of the animal welfare act, then they are obliged to care for that cat.

T: Mhm. So if I care for a cat [

I:] yeah, right

T: that's been abandoned, then it's, then it's my responsibility.

I: Yes, because there are like many cats who sit there and meow and they are wooing you to get food all of them, so they get some shrimp from someone, oh, and then they stay there, right.

T: Mm.

I: And then the people feel sorry for them, and then like now it's winter and then this cat's pitiful and, well, it continues. And most often it gets sorted out, 'cause really, when you make the decision to put a cat down, then, somehow, it works out anyway. Then people start saying, but yes, I know someone who can have, take this cat, so then [

T:] yes, you mean that it

I: [laugh] yes [

T:] it's like when the threat comes, then

I: yes, the threat of putting it to sleep, you know, it solves many things, it does. [laugh]

(Interview police officer)

The experienced police officer quoted above brings out the agency of cats: they make people feed them by hanging out in neighborhoods, meowing and being cute while begging for food. He goes on to say that the community around the homeless cat often seems to take care of the situation when faced with the reality that the cat will be killed. This alternative practice of collective cat "ownership"—in terms of informal responsibility, not in legal terms—is more common outside of Sweden and practiced in many communities, where a number of neighbors care for cat colonies by providing food and sometimes veterinary services (Miele forthcoming). This phenomenon suggests that the one-cat-one-owner model is not the only possibility. The point I am making is not intended to take a stance for or against these practices, but rather to indicate the possibility of alternative modes of multi-species urban politics and crowd control.

In sum, I have investigated how the feral cat, as free and ownerless, challenges the feline norm and is produced as morally bad, in need of correction through taming practices. However, this view is also charged from a more zoocentric perspective, where the ferality of cats is produced as equally precious and where the solution is feeding and neutering practices. The legal "one-cat-one-owner" model is challenged through cat/human collectives, by crowds of feeders and ferals. In line with Freud's concept of the "unheimliche," there is something disturbing about feral creatures. In his view, "unheimlich"—or uncanny—experiences and feelings appear when something that is hidden from our consciousness comes to the fore (Freud 1919 [2003]). Encountering the concealed or repressed creates certain negative responses, often fear, disgust, or even horror. It reminds us of our own shortcomings and fear of punishment for breaking social norms. In Freud's psychoanalytical eyes, the uncanny is essentially about taboos regarding sexuality and death, thus the uncanny is about the forbidden, which is produced as both repulsive and enticing. There is similarly something seductive about the feral figure, because it refuses to fit in to social orderings, thereby enabling a certain potential subversiveness.

Normalizing living—crowded shelters and sacred homes

It has been recognized that there are at least two ways of defining homeless people: either in terms of lacking a home—as houseless—or by their asocial behavior (Somerville 2013). Definitions are essential, as there is a link between them and explanations, notions of responsibility and interventions (Sahlin 1992: 51). In the struggle over definitions, different stakeholders and actors introduce and support different versions of reality. A municipality, for example, may be held responsible for caring for the homeless—both human and feline—and consequently may suffer financially. Thus, the existing figures cannot be seen as raw statistics mirroring reality, but as tools in the struggle over definitions (Swärd 1998). In my data, for example, shelter workers claimed higher numbers of homeless and feral cats than did the authorities. My argument here is not to say that cat home issues—including homelessness and shelters—are the same as for humans. There are some defining differences regarding blame and stigma, not to mention the moral discourse on poverty and individual responsibility. Few people would blame the cat for being immoral and living on the street as an individual choice. However, symbolically and institutionally, there are many overlaps.

Home is a place connected to putting down roots, identities are constructed in relation to where we live, where we belong (Duyvendak 2011). In the case of the homeless and feral cat, belongingness is obviously the dimension of home most strongly emphasized. The shelter movement, in turn, aims at bringing the homeless—in terms of without home, and sometimes the homeless in terms of antisocial behavior—to temporary housing at a shelter or in a foster home, while waiting for permanent "homing," for a new belonging. The animal shelter movement started out early in the late nineteenth century in the US and the UK (much later in Sweden). Its opponents argue that shelters make homelessness permanent through a certain culture. Irresponsible pet owners are excused when cats and dogs may be rehomed (Irvine 2003), and the home as a pure, sacred, and moral place can be saved.

Humanimal crowding, the central figure of this book, aims at capturing our understanding of the uncontrolled and the transformation of individuals in certain settings. It becomes rather obvious that, according to Swedish authorities and most shelter workers, the street is not a place for cats. Tame is the nodal point—a normal cat is tame and lives in a home. Still, symbolically it may be considered free and wandering, deceitful and "wild" in terms of untamed or even untamable.

How, then, should we understand the threat of the feral in terms of humanimal crowding and urban space, where feral populations appropriate the street and thus claim their right to the city? In many cities, people, mainly women, take it upon themselves to look after these cats (see Chapter 5). According to Swedish animal welfare law, cats must be owned by one person responsible for their wellbeing. Thus, colonies of feral cats are generally not accepted and, in practice, risk being eradicated by urban wildlife managers and animal welfare officers. Despite, or perhaps because of, these lethal legal interventions, neighbors often feed the cats and look after them in general, sometimes taking injured and sick ones to the veterinarian. Some of these caregivers even take care of neutering the cats, to make sure that they do not overpopulate (see also Davies 2011; Miele forthcoming). These "feral supporters" are produced as norm breakers, because they "dangerously transgress domestic norms" (Griffiths et al. 2000: 68). In line with the discussion of the previous chapter, I am inclined to think about these clusters of humans and cats in terms of trans-species urban crowds, threatening a certain naturalcultural order, where spatial boundaries between home and street, as well as between humans and animals, become blurred. The crowd and its historical connotations with wildness applies to notions of the untamed, feral cat and their feral supporters. The human end of the crowd is gendered female, which symbolically further strengthens the notion of the uncontrolled (see also Chapter 5 on "feline femininity").

Here, the notion of the "stranger" seems important. Feral cats are understood as outsiders or rather as strangers, who do not really belong, but who through their strange position reinforce normality: one cat, one owner, one home. The stranger, in Simmel's terms, is near and far, general and specific, in and out of the relationship.

> As a group member, rather, he is near and far at the same time, as is characteristic of relations founded only on generally human commonness. But between nearness and distance, there arises a specific tension when the consciousness that only the quite general is common, stresses that which is not common. In the case of the person who is a stranger to the country, the city, the race, etc., however, this non-common element is once more nothing individual, but merely the strangeness of origin, which is or could be common to many strangers. For this reason, strangers are not really conceived as individuals, but as strangers of a particular type: the element of distance is no less general in regard to them than the element of nearness.
>
> (Simmel 1950: 403)

Admittedly, Simmel is referring to human strangers, but the argument can easily be applied to a multi-species context (see also Jerolmack 2013). As stated in the quote, the stranger points at certain tensions in social relationships between the commonness—the norm—of the group, and that which breaks with it. Strangeness is socially and spatially produced, connected to emotions such as discomfort, fear, curiousness, and desire. The notion of cat-egorizations that I have highlighted

Figure 3.2 Street cat in China town, Manila, Philippines (Photo: Christer Barregren 2011)

in this chapter furthers this idea, in that it points out how important it is to discern the cat's origin. Is it common to "us," that is, originally from a home, a family, or is it from "the street" and thus non-common? Furthermore, as will be developed in Chapter 4, this ferality can spill over to the home, downgrading the "homeability"—the degree of homeliness—of both pets and place.

Some exploration is needed in the form of making a comparison between strangeness and displacement. The latter concept has different meanings in the literature, sometimes referring to the loosening of an attachment to a place, the disconnection—for example, through homelessness (Gieryn 2000: 482). The lost and the homeless cats discussed in this chapter could of course be considered displaced in this sense, like the missing cat from Colorado referred to above, whose disconnection was remedied through its return to the family and home. Other authors use the term for describing someone or something temporarily in the wrong place. In Mary Douglas's terminology, that which breaks with a certain cultural order is perceived as "matter out of place" and threatens to pollute the sacred realm (1997 [1966]). It could, however, also be conceived of as potentially magical and powerful. For example, in many cultures twins have been understood as breaking with a natural order, in which humans give birth to one child, only animals to more than one. But the interpretation of the out-of-place-ness of twins differs profoundly. In Swedish folklore, they were thought of as having magical powers, allowing them, for example, to protect the humanimal household against witchcraft and to cure illness (Holmberg 2005: 91). But, in areas of West Africa, their existences endangered the social order, and the normalizing strategy was to kill newborn twins (Douglas 1997 [1966]: 61). Sarah Ahmed (2000) writes that the stranger has the potential to embody the unknown as well as the known, but in racialized discourse is produced as a "body out of place," an outsider associated with danger. Places play important roles in accounting for meaning-making and management of homeless and feral cats, where the degree of homeability in relation to ferality is the decisive methodology of "cat-egorization." The stranger cat embodies a strangeness that, through its ambiguity, has the potential to radically question the meaning of places such as home and street. However, in law and urban homelessness management, it is reduced to being a body out of place, a threat to the neighborhood as well as to human health, and is normalized through crowd control technologies such as adoption practices, trap–neuter–release or euthanasia. Thus, sharing the lives of many involves a messy terrain where many actors come together and struggle over definitions and outcomes.

Part II

Humanimal transgressions

4   Verminizing

Making sense of urban animal hoarding

Swedish author and artist August Strindberg (1849–1912) is not primarily known for writing about animals in his fictional works, however they do occur in some of them. In the short novel *Tschandala* (1889 [2007]), the character named "the baroness" is depicted as a woman who, despite her assumed fortune and high rank, lives in filth together with numerous animals of many kinds. A visitor—Törner—is met by "repeated barks," "the subterranean barking of dogs," howlings, and mysterious shrieks (Strindberg, cited in Lönngren 2015). The visitor notes that, despite the baroness's outspoken love of her animals, of which she houses large numbers both in- and outdoors, they are obviously not cared for because they show signs of starvation and serious neglect. Thus, the visitor is puzzled over the fact that although the owner constantly stresses her

> boundless love for animals, the beasts were hardly given anything to eat. The horses chewed chaff with no oats, the cow licked moss or the mould on the walls or pulled down straw from the rotten thatch of the roof, and the poultry fought over dung. There was no bedding in the stables and the animals slept on their own excrement.
>
> (Strindberg, in Lönngren 2015)

Ann-Sofie Lönngren notes that this description can be viewed as an early testimony to a phenomenon that is currently called animal hoarding, and that, as such, it highlights contemporary norms on proper human/animal relations (2015).

A much brighter story is told about Kate Ward (1895–1979) from Camberley, England, and her stray dogs. Kate took care of the unwanted dogs of the community, and with the help of fellow citizens, she raised and cared for them. According to a local veterinarian who helped her with health services for many years, the dogs were quite healthy, and often lived long and "happy" lives. According to the story, Kate took care of more than 600 dogs during her lifetime, and this must have been hard work at times. What, then, was her reason for dedicating such a great amount of time and effort to stray dogs? "On the 12th November 1973, Kate was interviewed by a US TV journalist, George Montgomery, from NBC Evening News in America. He asked her why she liked dogs and her reply was that she was lonely and preferred them to humans" (BBC Surrey 2013).

Figure 4.1 "Stray dogs," Kate Ward, Camberley, UK (Terry Fincher/Getty Images)

Kate's reasons for caring for packs of dogs—love of animals who fill a self-chosen void of human relations—is fairly common when it comes to people's own explanations. But there are also other ways of making sense of the phenomenon. Evolutionarily, collecting in general is a behavior shared by many species, most famously the magpie. According to research, all people collect objects to a certain degree but, for some, collecting becomes a compulsive and often denied act. From 2013, an independent diagnosis labeled "hoarding syndrome," including the excessive collecting of animals, has been added to the growing number of psychological pathologies listed in the *Diagnostic and Statistical Manual of Mental Disorders* (*DSM-V*). This psychiatric term has also spilled over into the cultural sphere, where documentaries on animal hoarding and reality shows are broadcasted on primetime TV (see, e.g., Animal Planet's *Confessions*, or *Hoarders* on AETV). In these shows, people with animal hoarding behavior are portrayed as freaks who are unable to control their excessive and norm-breaking actions. For what appear to be strange reasons, neither can they part from the animals whom they obviously cannot care for, nor do they seem to care for human companions when the animals "take over" their homes and lives. It is as if they put the pack of animals before their own humanity. In general, the subjectivity and suffering of the hoarded animals are seldom acknowledged, as they are merely allowed supporting roles or portrayed as window-dressing in the mediated drama. The stories often follow the plot of a crime drama, as in the nationally rehearsed series of articles regarding a woman who hoarded wounded swans in her small apartment in Stockholm.

Figure 4.2 "Svankvinnan slår till igen" ("The Swan Lady strikes again") (Aftonbladet 2007)

Although stories such as the introductory ones above suggest a historical continuity of animal eccentrics, newspapers and other media report that animal hoarding has become more common and that it is a growing urban problem (Arluke et al. 2002). This chapter sets out to make sense of the phenomenon of excessive pet keeping that breaks with modern norms and policies in an urban context. Much like Strindberg's disputed character and the storied life of Kate Ward, overwhelming pet relations and the conditions of the home clash with norms and affect the ways in which both humans and animals are perceived. In comparison to the cases above, however, the contemporary legal context is much more strictly regulative with regard to human/animal relations. In the following, I explore the double meanings of "sense-making"—in terms of both understanding and explaining the phenomenon of animal hoarding and employing sensuous resources—through policy, practice, and discursive work. In addition, I look at the non-human dimensions and agencies of home making and governance: how practices are shaped through material, non-material, and non-human components (Jacobs and Gabriel 2013: 217). I would like to add here that the agency of hoarded animals is one such neglected and crucial component. Important to point out, it is the urban and suburban home that is in focus, with its specific spatial context: the neighborhood, the garden, the stairway and, not least, the proximity of neighbors. Most often, it is the people living next door who report to the authorities about disturbances in terms of noise and smell, signs of excessive pet keeping and potential neglect, arguably signaled by the animals themselves.

When it comes to sense-making, first of all I am interested in how animal hoarding is made sense of, i.e. understood and explained. How do psychologists, social scientists, and TV productions account for the strangeness of the phenomenon? How do people who work with these issues—urban animal management authorities—explain the norm-breaking behavior of hoarding? And how do animal hoarders themselves make sense of their behavior? In ethnomethodology, a central concern is the way in which members of a community develop methods to understand, navigate, and manage social situations (Garfinkel 1967). Thus, members' framings and explanations of social phenomena are crucial to their practices. In real situations, understanding and practice are interwoven, inseparable dimensions of social interaction. However, analytically, making this separation explicit may be fruitful.

Therefore, second, I am interested in how actors assess animal hoarding in the context of the explanations and policies that are discursively available. When people cannot care for their animals properly, or their excessive pet keeping is viewed as a disturbance to neighbors, the animal welfare authorities may get involved, sometimes assisted by the police. And like the officials carrying out farm animal welfare assessments, these urban assessors use a combination of objective data (numbers, temperature, air climate, etc.) and subjective impressions of the animals (fur texture, smell, movements), along with their feeling for the place. Assessors thus need to rely on the "look and feel" of the whole context (Roe et al. 2011: 69), employing their whole sensory register to produce knowledge. Here, sense-making in terms of sensing the environment is put to the fore. I ask how the authorities know when complaints are justified—according to the law—and how sensuous, environmental, and animal agential resources are used to assess the place and come to a decision about how to proceed. I am thus concerned with "sensuous governance" (Holmberg 2014b), a term that refers to the ways in which this sensing as a way of knowing gets involved in asymmetrical power relations and the method of assessment: knowledge and power go hand in hand. The framework of sensuous governance allows me to focus on the ways in which authorities use and record their senses of the emplaced situation—their visual, olfactory and auditory impressions—in order to make a judgment of animal hoarding in urban homes. I address this by attending to the close interrelation of emplaced knowledge and power in the emplaced, triadic interaction between authorities, owners, and animals. Towards the end of the chapter, I connect the analyses with the broader framework of humanimal crowding in zoocities, and argue that hoarding can be viewed as a verminizing phenomenon: too many, too unruly, too dirty. Through the verminizing of the home, it spills over to both the non-human and human inhabitants. I will discuss how hoarded crowds and crowding and the meaning it takes through the spatial context—the urban home and degrees of homeability—can be understood through the lens of trans-species sensuous emplacement.

Animal hoarding—a stigmatic phenomenon

We have witnessed changed norms concerning what it means to be human and act humanely in relation to other animals, and policies regarding pet keeping and animal welfare have been sharpened worldwide (Franklin 1999, 2006). Kate Ward and her band of dogs would probably not be tolerated today, and her behavior would risk being pathologized. When the amount and demands of animals exceed certain limits, and the everyday lives of the human(s) are affected, the phenomenon is currently called "animal hoarding": "Animal hoarding is keeping a higher-than-usual number of animals as domestic pets without having the ability to properly house or care for them, while at the same time denying this inability" (*Wikipedia* 2013). This common-sense definition highlights how animal hoarding is not so much defined by particular species but is about keeping large amounts of animals *as* pets. Thus, farmers and broiler producers are excluded, while people collecting "wild" animals and keeping them in the home, are included. Urban animal hoarding is characterized by the crowding of human and non-human bodies in housing not fit for as many inhabitants. The hoarded animals often suffer badly in these environments and, for example, display stress-related behavior problems, diseases due to lack of proper vaccination practices and veterinary care, reproduction-related conditions following excessive breeding, fur problems, and starvation-related illnesses, and dead animals are quite often found during inspections. In addition, the hoarders often fail to provide care for themselves and dependent others, manifested through psychological and somatic illness. The suffering of animal victims and their human hoarders are often co-dependent (Nathanson 2009). Besides suffering, there are also substantial costs associated with hoarding. Police and animal welfare visits and handling, veterinary care and euthanasia, animal shelter costs, litigation and clean-up can easily become very costly, and the subject is often not able to pay (Patronek 2008). Thus, animal hoarding is a complex human/animal public health problem (Castrodale et al. 2010), a social problem, and a housing problem.

Donna Haraway states that social science scholars need to "stay with the trouble" in order to help create more ethical and "response-able" relationships with other animals (2010). There are no clear-cut answers, not even clear-cut questions, when it comes to issues regarding human/animal relations. Instead, they are almost always—perhaps more than other social relations—paradoxical and conflicting, complex and confusing. When it comes to animal hoarding, the relations are perhaps more confusing than ever. My task here is not to suggest that the authorities are "wrong," or that the owners are "cruel," that they love or do not love their animals. Instead, I want to point at the importance of attending in detail to the qualitative aspects of human/animal interaction and relations in their particular settings, and the role that senses and embodied impressions play in accounting for these encounters.

The chapter starts with an in-depth study of behavioral scientific accounts of animal hoarding by looking at the classification and construction of pathology. Moreover, analyses of representations of hoarding in media reports and US reality shows will be used to illustrate the questions posed in the chapter. In addition to the twenty-one interviews performed with Swedish animal police, animal welfare officers, cat shelter workers, and people who have been subjected to inspections made by the authorities—so-called animal hoarders—the chapter is based on detailed and anonymized animal welfare complaints, protocols and decisions from twenty cases, derived from two of the county administrative boards. Each protocol consists of both standardized sections to be filled out by the officers and more qualitative records, describing the situation and the communication going on, and the official decision letter, sent to the person under investigation. In some cases, the documentation consists of extensive protocols from a number of visits, while in others the information is thinner.

Empirical social science studies engaged directly with animal hoarders are sparse, probably for good reason. It is a stigmatized group and "hoarding" is a stigmatizing concept. A stigma is, using Erwin Goffman's words, a label that is used to identify a deviant group in terms of visual and other characteristics, on the one hand, and that contributes to the self-identification of that group, on the other (Goffman 1990 [1971]). Moreover, the stigmatization process functions so as to strengthen the norms and normality of the rest of society. Thus, categorization is not an "innocent" practice: knowledge and power go hand in hand (Foucault 1998). I will be addressing both the construction of deviance and the effect of stigmatization processes—the strengthening of norms—by examining explanations and assessment of animal hoarding.

Explaining animal hoarding

While cities are becoming more and more crowded due to urbanization, some people crowd their homes with animals, and quite often also with "stuff" (Frost and Steketee 2010). In the behavioral science literature, research on animal hoarding and research on object hoarding are typically separate. The *DSM-V* manual from 2013 includes "hoarding disorder" as a diagnosis, but with an unclear pathology. Several explanatory models exist side by side, and the prevalence of co-morbidity is also acknowledged: one individual's disorder may well have several causes. Recent studies have revealed that between 2 and 5 percent of the population show some kind of hoarding behavior that causes them distress and interferes with "their ability to live" (Frost and Steketee 2010: 9). Among this category, a small proportion also collect living animals. Frost et al. (2011) bring studies of object and animal hoarders together, suggesting that hoarders are not easily defined and that there is still little known about the background of animal hoarding. Experts are acutely aware that this group is very hard to treat, because hoarders seldom think there is anything wrong with them and because they—according to the authors—often have severe psychological and social problems (Patronek 2001; Frost and Steketee 2010). It has also been suggested that hoarding disorder is hereditary, and genetic studies have been performed. Moreover, delusion (Lockwood 1994), dementia (Patronek 1999), and addiction (Lockwood 1994) have been suggested as frames for understanding and explaining animal hoarding.

A psychological disorder

One very common explanatory model ties animal hoarding to obsessive com-pulsive disorder (OCD), where hoarders are characterized by an "overwhelming sense of responsibility for preventing imagined harm to animals" and as going through unrealistic rituals in order to prevent such harm (Frost 2000: 27). This notion is challenged by others, claiming that when it comes to "hoarding of sen-tient beings," there is not much evidence of compulsive, repetitive behavior, and that patients seldom respond to treatments used for OCD (Nathanson 2009: 317). Instead, another common explanation—the attachment model—is suggested as an alternative, where the pet becomes a "fix," a method to fill a void of "unsolved early losses" (Nathanson 2009: 317). It may also be that "the individual suffers from early developmental deprivation of parental attachment," for example as a result of having had physically and sexually abusive caregivers. Because the adult with attachment problems cannot establish close human relationships, the unconditional love given to and received from pets is preferred (Nathanson 2009: 317). Interestingly, none of these studies is based on systematic clinical studies of animal hoarders, which confirms that this is a difficult category to study. To fill this gap, the Hoarding of Animals Research Consortium (HARC) was formed and started studying animal hoarders in more systematic ways (Frost 2000: 26). An early study of nine women, most over forty years of age, suggests that,

> characteristics included the beliefs that they had special abilities to com-municate or empathize with animals, that animal control officers failed to recognize the care the interviewees give to their animals and that saving ani-mals was their life's mission. Typically, animals played significant roles in their childhoods, which were often characterized by chaotic, inconsistent and unstable parenting.
>
> (Frost 2000: 26)

Seen from the hoarder's point of view, if any explanation that he/she provides is interpreted as delusional, or as leading to other explanations that confirm pathol-ogy, then his/her motivation to participate is perhaps not particularly strong. In Chapter 5, I will discuss the moral and gendered dimensions of these accounts, where explanations often portray women as traumatized victims of, for example, sexual abuse and animals as substitutes for loving human relations. The behav-ioral scientific explanations tend to obscure the fact that the hoarded animals are also victims of abuse. Moreover, it becomes clear that human/human relations are the norm for a healthy, social life—people should not engage too much with other animals.

A social problem

Social science studies of animal hoarding are quite rare. Sociologist Arnold Arluke is one exception, as he and Celeste Killeen wrote about a specific and extreme case of hoarding in the US: Barbara Ericksson and her 552 dogs (2009). A striking message within the book is how authorities failed to handle the situation: when the situation got hot in one place, Barbara took her dogs and moved to another town. One key to understanding how this could go on for so many years is that, in rural communities, such eccentricities are more accepted (2009: 154). In the end, although the majority of the dogs were rehomed, a large proportion were in such bad condition that they had to be euthanized. The authors state that, given the lack of knowledge in the area of animal hoarding, and the lack of collaboration between institutions such as health care, social care, and animal welfare, success-ful prevention and management is rare (Arluke and Killeen 2009: 179).

When the media report on hoarding the attention is often quite intense and dramatic, but often fails to acknowledge the suffering of the animals (Arluke et al. 2002). Media coverage often follows the logics of the crime story but, contrary to the normal script, seldom discusses the victims (Arluke et al. 2002). The effect of this neglect is that, despite the fact that these reports are rather frequent, the animals become highly invisible. Instead, neighbors, lawyers, experts and, more rarely, the owners themselves get to speak. Based on media accounts, sociologists have developed different categories of animal hoarders (Arluke et al. 2002). They are portrayed either (1) as criminals, as animal abusers who will face punishment, (2) as psychologically ill, as mentally unstable, or (3) as pitiful, pathetic less-than-human creatures (Arluke et al. 2002). With this backdrop, it is not surprising that hoarders use different strategies to "de-stigmatize" themselves by explain-ing or justifying their deviant behavior (Vaca-Guzman and Arluke 2005: 340). Arluke and colleagues touch upon an interesting aspect when they state that, at the societal level, the increased prevalence of homeless cats and dogs provides one possible structural explanation for hoarding behavior (Arluke and Killeen 2009: 178). Hoarders often claim that they have an important role to play in caring for abandoned strays, and sometimes they are actually categorized as rescue hoarders.

Gary Patronek (2008) underscores that animal hoarders are very difficult to study, because they—besides having mental problems—often suffer from various health problems due to poor sanitation, and often fail to acknowledge their respon-sibility. Some studies, like Patronek's, have dealt with demographics and preva-lence, showing somewhere between 0.25 and 0.80 reported cases per 100,000 people in the US (1999: 228). The findings show that hoarders, unlike other types of animal abuser, are typically female. They are often interested in one or two species (most often dogs and/or cats, but also other pets like rats, mice, rabbits, or snakes), live alone, and are over the age of forty. In Patronek's study, based on a review of questionnaires from local US authorities, the majority of homes were heavily cluttered with garbage and were unsanitary (urine and feces everywhere), and the animals were in poor health, anti-social or even dead (1999: 226). Most of the cases were reported by neighbors in urban or suburban areas, and there may be a higher rate of reporting in the city than in rural areas. In Sweden, there are limi-tations on how many animals one is allowed to have in a city house or apartment (no more than nine), and many private rentals have much stricter limitations (Dog and Cat Supervision Act 2008: §9). However, it is important to note that it is not the numbers that define the hoarder, although a large number is a strong indicator. Instead, the problem is that the number "overwhelms the ability of the hoarder to provide acceptable care" (Patronek 2008: 220). This definition differs in focus from the one referred to above, such that the decisive criterion of the latter definition is that the hoarder cannot care properly for her/himself or for dependent others living in the house or apartment.

It is somewhat striking that none of the studies reviewed above is concerned with analyzing the construction of hoarders as pathological, as deviant. The data are accepted at face value, and thus the studies contribute to the production of deviance and the focus on the social problem. Another noteworthy lack of perspective concerns the human/animal relations taking place, considering how the norm-breaking interaction with animals contributes to the notion of deviant human behavior.

Loving animals

In the literature, animal hoarding is thus explained in various ways; there is no single etiology. Although the phenomenon now has one separate correlation in the psychiatric *DSM-V* manual, at least three types are mentioned in the somewhat sparse literature: overwhelmed caregivers, rescue hoarders, and exploiters (the latter with no positive feelings for the animals) (Arluke and Killeen 2009; Frost and Steketee 2010). Excluding the third category, Frost and Steketee state, "At the core of most animal hoarding cases is a special feeling for animals, a sense of connection that was hard for the people we interviewed to articulate. Pamela described it as 'pure love' while others we interviewed described it as 'beyond love' and uncomplicated by less worthy human emotions" (2010: 129). It is suggested that the love of animals differs from and replaces emotions directed at humans, and that this is part of the pathology.

The explanation that animal hoarders really love animals too much was also by far the most common one in my interview data. For example, when asked about the reasons for taking on too many pets, one police officer states:

I: Because it, and, these ladies, or I shouldn't say ladies but . . . who have this mania and collect cats and they become more and more, that's people right, they love their cats. It's not as if they don't give a shit. They love their cats.

(Interview police officer)

This love is not unproblematic, as in the following quote from an animal welfare inspector—it tends to go too far:

T: What's your experience? How would you explain it [hoarding]?
I: There are those who say that, "well, I like animals so much. I like animals so much. And then I saw this rat here and that cat there and that dog on the

internet . . . I like animals so much," but, in the end, you have reached your limit, right.

(Interview animal welfare inspector)

According to the quotes above, at first the hoarders have good intentions and want to rescue stray and unwanted animals because they genuinely care for them. The problem arises when they take on too many. Again, numbers seem at first glance to be decisive, here used as an explanation: in the end they reach the limit. However, it seems to be the love itself that is the problem, a love that is norm breaking, as in the following quote from a cat shelter worker:

I: Well, right, I think basically it's about you, you like the animal so much. You like it so much, and perhaps you have been excluded and in, the cats are your best friends, they are the ones who are always there and can speak with you and are there for you and everything. Well, perhaps you don't have so many social contacts. And then it gradually degenerates. And then you . . . instead of being a really fine friend to an animal, you become an abuser of animals. [. . .] From liking animals a great deal to . . . abusing them, it's not so far.

(Interview cat shelter worker)

The shelter worker above does not seem surprised over the fact that love can go wrong. Moreover, a supposed lack of human social contacts helps in explaining why the animals become so important. Even though far from all people who claim to love animals, including those labeled as hoarders, lack human relations, their outspoken commitment to other animals challenges societal norms and conventional identity categories.

As an illustrative example, one episode from the Animal Planet series tellingly entitled *Confessions* depicts an American couple, Shelley and her husband, and their "family of 67" (Animal Planet 2012). This particular episode illustrates how hoarding, housing, and humanness become co-constructed: the excessive love of animals, together with a perceived lack of (proper) human social relations, is presented as a loss of control and low degree of "homeability" (see Chapter 3). Shelley and her husband Chris are said to live in a house that belongs to the cats, instead of the other way around, as it should be. The show follows a plot typical of this genre. First, the situation is presented by the actors themselves, often in a quite positive light. Shelley repeatedly states that she loves her cats, her "kitties," every one of them, and that they all have names. When new "babies" are born, it is as if she becomes a mother too. Plenty of work goes into organization of the living space, where one room is for sleeping, one is for eating, and one is the "potty room." Chris continuously rebuilds the house for the cats, sawing holes in the walls and constructing a feline-centered home. Much like Julie-Anne Smith in her article, "Beyond dominance and affection: living with rabbits in post-humanist households" (2003), the house is rebuilt in order to fit the human/animal cohabitation. However, when close others like friends and family get to speak, they reveal a very different story. Shelley's relatives worry that the couple believes "this is normal, but it's not." It becomes clear that human/human relations are the norm for a healthy, social life—people should not engage too much with other animals—and that this prescribed life is saturated with norms regarding gender, age, class, and sexuality. The camera vision underscores the situation by letting the viewer peek beneath the surface, revealing cat feces, and ripped furniture, walls and textiles. Later, Shelley, according to the general script of the series, "confesses" that she has a problem, and experts comes in to help relieve her and her husband from the overwhelming burden of the cats. Others who appear in the reality show make similar confessions: they love their animals, they are their babies, they cannot live without their animals, and so on. Confronted with family members and experts, they will in the end admit that the situation has gone awry. For example, Mike, who lives in a totally worn-down house full of cat feces—the ultimate sign of decay—says with tears in his eyes that he feels "ashamed, embarrassed and weak . . . I am mad" (Animal Planet 2013a). Being pointed at as a deviant person, thereby stigmatized, also involves feelings of guilt, shame, and inferiority.

How do animal hoarders account for their behaviors in interviews? I asked one upper-middle-aged woman how she had the time and commitment to tend to the twenty-five dogs she had, prior to the unannounced inspection from the welfare authorities that changed her life.

T: How do you find [laugh] the time?

. . .

I: Well, yeah, yeah, but it's so fun.
T: Huh.

[pause]

I: It's what life is about [laugh].

(Interview dog owner)

The interviewee is deeply devoted to a particular dog breed, the small and furry Yorkshire terrier. She admits that she, out of ignorance, kept too many of them, but states again and again that she did not harm them: they were her life. Later in the interview with the same dog owner, we come to the explanations in more detail (what are the driving forces) and I also challenge the supposed ignorance: What did she do wrong?

T: What if we were to talk a bit about your eng, your engagement then for [
I:] yes.
T: for animals and particularly for small dogs [
I:] yes, yes.
T: you can say, Yorkshire terrier. What, what, how, how do you view your role as a dog owner today. Or as . . .

I: Ex dog owner [

T:] Mm.

I: Yes . . . well I have [pause] I've eh . . . once made a mistake. But I've been
incredibly kind to my dogs and I've loved my dogs. I haven't . . . been one
of those Paris Hilton types but I have been out with them in nature and have
loved to watch them run in the woods and . . . it's my recreation and then I've
photographed them.

T: Mm, mm. . . . What is, what's your driving force, what is it that makes you
love these dogs?

I: Well . . . it's simply my family. I was raised, I was . . . raised [laugh] as a
Yorkshire terrier.

T: [laugh]

I: My mother brought me up as a Yorkshire terrier [laugh].

(Interview dog owner)

This lengthy quote illustrates how difficult it is to ask and speak about the
emotionally charged accusation of animal hoarding, and the consequences of
this behavior: a life without dogs. I, as the interviewer, hesitate and repeatedly
rehearse statements ("what, what, how, how do you view your role as a dog owner
today. Or as . . . "), while the interviewee similarly struggles with formulating
that she did make a mistake ("Yes . . . well I have [pause] I've eh . . . once made
a mistake"), then covering that mistake by again claiming that love was the basic
emotion. These sequential tensions are then relieved through joint laughter at the
interviewee's statement that she was brought up as a Yorkshire terrier. She later
states that she did not have any children, but that "they were like . . . my children . . . my Yorkshire terriers." Within the psychological taxonomy, this dog
owner would possibly fit in to the category "overwhelmed caregiver," buying but
not breeding dogs predominantly of a particular breed. However, with a family of
25 terriers, this "mother" does not succeed in providing care for the dogs, in fact
the welfare inspection report depicts a horrendous situation and, while some of the
dogs were rehomed, some were also euthanized.

Another interviewed pet owner had also received unannounced calls from the
welfare authorities. Neighbors had been complaining about the noise and smell
from, and possible mistreatment of, innumerable cats. This woman, also middle
aged, has a rather different story. She "confesses" that she did have too many cats
and claims that the situation got out of hand due to financial stress and overwork.
According to psychologists, she would fit in to the category of rescue hoarder—
she could not resist helping out cats in need. From the very beginning of the
interview, she frames herself as an animal lover:

T: I thought I'd ask about you personally, how old you are, where you come
from.

I: Mm. Well, I am XX. Come from XX, and am a total animal freak [laugh] I
like all animals except ticks. Work at a vet's clinic.

T: Okay.

I: And I've started up two cat shelters . . . And I'm involved in Animal Rights Sweden. So I really have a passion for animals.

T: How did that interest begin? [pause]

I: Well, how did that interest begin? That's a good question, I really don't know that. It's probably something I was born with . . . Yes. So, you know, I've always been interested in animals, all the time.

(Interview cat owner)

This lifelong interest in and love of animals is not discriminatory, she has cared for rabbits, horses, and hedgehogs: "it doesn't really matter what kind of animal it is." Now she has a dog that she takes care of. When I ask about the rest of the family, I am told that she does not have any children, and that "animals are my children."

Loving animals and at the same time abusing them. How can we make sense of this discrepancy in a way that moves beyond pathologizing the relationship? When struggling to understand this figure of animal love in the context of "passive cruelty" (Vaca-Guzman and Arluke 2005), I argue that a sociological understanding of animal love in the context of hoarding has to move beyond OCD. In addition to understanding normative human/animal relations, it has to account for the simultaneous love of animals and the (sometimes lethal) harm often inflicted on loved ones, human or non-human. Dependency and domination are not exclusionary, but intertwined dimensions of care (Kittay 1999). As such, a theoretical analysis of love, dependency, and care from the sensorial embodied perspective accounts for the dialectics of instrumentalization and exploitation of, and care for, animals (Holmberg 2011b: 160–161). The category "animal hoarder" is connected to a stigma; it is norm breaking in numerous ways and the people in question often feel threatened, ashamed, and upset when confronted by the authorities. Perhaps explaining their behavior by referring to something essentially positive, that they really love animals, is a way of saving face and restoring a "spoiled identity" (Goffman 1990 [1971]; see also Arluke and Killeen 2009: 192).

But it is also a love that is not "proper"—if one cares too much for animals and lives too closely with them, there is a risk of becoming less human, being dehumanized or even animalized, as will be explored further in the following. However, it can also work as a liberating refusal of categories. Kathy Rudy describes her own identity in terms of a dog lover, as one that cannot be named: "It's not so much that I am no longer a lesbian, then, it's that the binary of gay and straight no longer has anything to do with me. My preference today is canine" (2011: 41). The animal owners that I quote above refuse to refer to themselves as childless: like Shelley, the mother of 67 cats, the animals are their children. But it is a gendered relationship, men with hoarding tendencies rarely say that animals are their babies, they are rather framed as their friends or in other ways conceptualized as being in need (see Chapter 5).

Out of control

Although most of the informants explain the phenomenon of animal hoarding as love that has gone awry, other explanations are also discursively available.

I: Well, there are . . . there are the . . . well people who have animals and they can be, farmers for example who, start feeling psychologically ill like anyone else, they are confronted with a crisis [

T:] Mm.

I: in their life and just can't take care of their animals. And there is nothing strange about that really, that people end up in crisis situations. But then there are, I think, those who aren't feeling well from the beginning and start collecting and so on, and then they become more and more . . . And they shouldn't be allowed to have animals, I think.

T: Right

I: But then we often get this situation, with 20 cats in a small house and no cat sand and no food and . . . all that. And that is, well, terrible.

(Interview animal police)

This police officer contrasts the case of a psychologically unstable animal hoarder whose behavior goes awry ("not feeling well from the beginning," "start collecting . . . more and more"), with a more crisis-oriented understanding. According to interviews, the crises may be due to emotional (loss of loved ones), financial (loss of income), or health (severe illness, hospitalization) problems. Consider, for instance, the following excerpt from an inspection report: "XX does not know how many cats she has indoors or outdoors. She is ill and has not managed to keep clean . . . XX lost control over her cat keeping long ago. The cats are confined to a totally unacceptable environment that is detrimental to the health of both animals and humans" (Inspection report municipality X 2008). Illness and the woman's inability to keep the house tidy are read as signs of lack of control and thus have explanatory value. Naturally, not all cases come to the authorities' attention; sometimes help is sought from cat and dog shelters.

I: Well, most often it's neighbors kind of, or others [

T:] Mm.

I: or friends you know who sort of observe that this is gonna turn out bad and the old lady spends too, all her money on cat food, you know, and that's not possible.

T: Mm.

I: Eh and . . . can't keep it tidy and, well, like that. It [

T:] Mm.

I: it becomes weird, it, old, old cat poop in the corners [

T:] Mm, mm.

I: and it, like that. Eh and the cats aren't properly socialized either.

(Interview cat shelter worker)

Figure 4.3 Cat in bed (Photo: County Administrative Board, Skåne)

Friends and neighbors in the urban community observe what is going on when things are getting out of hand, and may contact the local shelter. Spending all your money on cat food and tolerating cat poop are seen as evidence of loss of control. Control and lack of control are also central themes in the already mentioned reality show *Confessions* (Animal Planet 2013b), where we get to know about Johnnie, who started collecting dogs after his father passed away. He now has 13 dogs, and they have, according to Johnnie's sisters and other family, taken over his house. "He's been run out of his home, by his dogs basically, and he lived in his car for a while before anybody found out." The dogs have destroyed the house, it is said, "it's beyond bad, it's not livable." Meanwhile, the viewer gets to see a house torn to pieces, cluttered, and full of feces. "He'll feed his dogs before he feeds himself, and that's sad." The fact that Johnnie seems to put the dogs before himself reverses the species hierarchy and upsets viewers.

Frost and Steketee write that an "odd feature we observed was that the hoarders became more animal-like in their daily habits over time" (2010: 129). In their view, when the animals are allowed to "take over" the homes, the human somehow becomes less human and more "animal-like." This is a common theme in reality shows about animal hoarding. In one episode of *Hoarders* (AETV 2012), for example, featuring Kathy and Gary, who are living with many rabbits, the therapist says: "It seems like the rabbits live here, and you guys are the pets?" This indicates a confusion of species hierarchies, and one could say that interventions by authorities aim at re-establishing boundaries (who is the human, who is the pet), but also the proper place for pets (Smith 2003). Verminizing involves the dimensions of number, space, motion, and affect. When animals become verminized, the human(s) is also conceptualized as being out of control. Through their behavior, humans become animalized, represented as less than human. One undisputed sign of control loss is the prevalence of animal feces and the handling of it:

> The dogs relieve themselves on the floor and the dog owner XX picks up the dog feces with her bare hands. . . . There is dog feces behind the sofa, the armchairs and on the carpet. The sofa is soaked with dog urine. Also in this room, XX picks up dog feces with her bare hands. There are traces of dog feces all over the (parquet) floor.
>
> (Inspection report municipality X 2007)

In the next section, I will return to the issue of (lack of) tidiness and animal feces as central themes in animal hoarding documentation. Here, I focus on the ways in which hoarders are understood as having lost control due to their close contact with animal excrement. Feces and urine should be handled at a distance, and as part of our civilization process, this separation between the human body and its excrements is essential (Elias 2000; Waltner-Toews 2013). Picking up feces with your bare hands is so utterly uncivilized, and in the above documentation, it is unquestionably a valid sign of loss of control. Rationality and control, valorized human capacities, seem to have disappeared.

Even though the human actor is portrayed as losing control, it does not mean that the animals automatically are "in control," although this seems to be the plot exemplified above. On the contrary, the visual representations in the programs, as well as protocols and other evidence recorded by the police and animal welfare authorities, show the lack of control: lack of food and water, lack of space for exercise and clean indoor air, lack of health, veterinary care, birth control, etc. So how are we to conceptualize this apparent discrepancy? This is where the assessment process comes in.

Detecting and recording animal hoarding

Animal presences in urban space are ubiquitous and often contribute to the identity of places (Philo and Wilbert 2000). But the involvement of animals and the human/animal relations taking place are often contested (Holmberg 2011a).

The contact zones of multi-species encounters, where species meet and become "messmates" (Haraway 2008a), are of course numerous and heterogeneous. In the present chapter, it is the urban and suburban home of people and their animals. More importantly, perhaps, it is a home where officials enter uninvited in order to "perform animal welfare" in the name of the law (Roe et al. 2011). According to the law, on the one hand, humans own animals as property, which from an animal studies perspective is a complicated stance, reinforcing the dichotomies and power relations problematized (Miller 2011). On the other hand, animal welfare laws worldwide, state that the owner is inclined to care for the animals properly and may lose their animals if they fail. Thus, ownership may be challenged in the context of hoarding. But how is this proper treatment assessed? When encountering the place and assessing the interaction between the humans and animals present, authorities such as animal welfare officials and police officers need to make use of their various senses to get a feeling for the place. When encountering the place, including the humans and animals present, these officials need to make use of their various senses to get a feeling for it. Moreover, when encountering non-human animals, verbal interaction is not primary (although some studies point at the importance of symbolic interaction, see Sanders 1999). Instead, affects and the sensorium are important embodied communication tools. Actors use their visual, auditory, and olfactory capacities, as well as their sense of touch, to record and understand what is going on—in short, they use sensuous methods to produce knowledge about their social multi-species environment.

The trajectory of a "senseological," or sense-sensitive, approach goes back at least to Georg Simmel, who argues that attending to the details in "sensory impressions" can deepen and further our sociological understanding (Simmel et al. 1997: 110). Michel Serres talks in a similar fashion about the "topology of senses" in an attempt to understand how sensing is always a matter of engaging with a multitude of impressions that are intermingled but hierarchical: we do not encounter the world through all the senses in equal amounts, but the quality of impressions shifts with context as well as personal experience (Serres 2008). David Howes uses the concept of "intersensoriality" to highlight that sensual perceptions mediate one another in complex ways (2005: 9). It is important to note, however, that we experience sensual impressions as separate, operating more or less "in relation to each other" (Pink 2011: 266). Regarding cross-species bodily encounters, Eva Hayward invents the trope of "fingeryeyes" to account for the ways in which cupcorals and scientists at a marine lab encounter one another "to name the synaesthetic quality of materialized sensation" (2010: 580). According to her, perceptions of other animals are generated by the interspecies encounters themselves.

Furthermore, senses are not strictly personal. Even though they are experienced by individuals, senses can be understood as constructed through cultural systems, mediated, and regulated through norms and belief systems (Howes 2005: 5). This approach, in line with Simmel's and Hayward's above, enables us to understand sensing as being integrated into knowledge production: "The senses are gateways of knowledge, instruments of power, sources of pleasure and pain . . . " (Howes 2005: back cover). However, although often acknowledged, this connection between senses and power—an aspect of what I have labeled "sensuous governance"—is seldom elaborated on. Thus, sensing can be viewed as technologies of power, and "sensuous governance" as reinforcing hegemonic norms of proper living and human/animal relations based on sensually constructed knowledge (Holmberg 2014b).

Living with many animals often creates controversies with neighbors, particularly in an urban context. The smell and noise of crowds of animals, mainly dogs and cats but also other pets or even wild animals such as birds, sometimes cause people to react. Complaints are filed, often anonymously, with the local authorities. This excerpt is from a concerned neighbor, writing in the local newspaper:

> Someone living in our neighborhood has a lot of cats and it smells from far away. Some days we cannot sit in our gardens . . . This is a problem for both the cats and the neighbors. How do we continue? Should one file a complaint about the cat owner? . . . This has to come to an end!
> "Upset neighbor"
>
> *(Arbetarbladet* 2010)

In Sweden, complaints as called for in the quote, are filed either with the police, who handle disturbances, or directly with the authority that deals with animal abuse and neglect, the local County Administrative Board, which then takes action in line with what is stipulated by the law. The Swedish Animal Welfare Act states that, "animals shall be treated well and be protected towards unnecessary suffering and disease" (1988: 534, 2§; author's translation). This more general frame (i.e. animals are not treated well) in policy and practice boils down to external measurements such as individual space and bedding, air quality, availability of food and water, along with health parameters such as fur quality, nutritional condition, behavior, and general physical formation. Sometimes the case is given less priority, for example when the evidence is not strong or the complainant is not considered trustworthy. Sometimes the owner is notified by mail. And sometimes the animal welfare officers make unannounced calls to check the conditions for themselves. If the inspectors visit the home and find that the case involves violation of the law, they can roughly choose between letting the animal owner/caretaker keep the animals and correct the conditions or having the animals removed. But how do the animal welfare inspectors come to a decision? What priorities do they have, and what cues do they use to help them make a decision?

The first sniff

As mentioned above, the animal welfare inspectors make unannounced house calls in an attempt to detect suspected mistreatment of animals. But sensing the place starts even before they enter the house/apartment:

> The animal welfare inspectors XX and XX arrive to the address in question. From the dwelling, barking from a larger amount of dogs is heard all the way

out in the street. Lighting is on at the upper level and the kitchen. When the animal welfare inspectors get closer to the landing, a very heavy stench of dog excrement and urine is sensed. There is dog excrement on the threshold of the front door. No one is at home. [. . .] When XX opens the door, the animal welfare inspectors were met by an awful stench. The ammonia odor is so heavy, it stings their eyes and the animal welfare inspectors experience nausea. The dogs bark continuously and the noise level is so high, it's difficult to have a normal conversation.

(Inspection Report County Administrative Board X 2007)

In this case, the noise from a crowd of dogs, the smell of excrement and urine (and the sight of it), together with sensations of stinging and nausea upon entering the house, give cues as to whether there is mistreatment going on. Starting from the first sight, sniff and sound, the officers begin to organize their impressions by way of recording them. This illustrates what Steven Feld (2005) talks about as a "sensuous epistemology of environments," in which not only senses are placed, but places are also sensed. Sensing becomes an essential tool in creating knowledge about a particular place. Perhaps needless to say, the visits are not appreciated by the human inhabitants:

When the County Administrative Board was about to enter the small property at approximately 17.00 hours, XX stated that there were 12 large cats and 12 small cats in the house. XX and XX refused to open the door. One police officer entered the house through a window at the front of the house. He opened the door between the kitchen and the hall to let the County Administrative Board inspectors in.

(Inspection Report County Administrative Board X 2010)

The communication, or lack of communication, between the owners of the animals and the authorities is often documented. Inhabitants may refuse to open, as in the case above, or pretend that they are not at home. Challenging the "epistemology of the environments" that the welfare inspectors provide, cat and dog owners themselves often do not approve of the unannounced visit and the judgments made, and complain over the procedure:

I: And then they rushed in . . . had no contact or discussion with me . . . but were talking over my head, some sort of monologue. And I tried to say that this is something I can deal with on my own, having so many friends who can, can . . . that I can transfer these dogs to. And they are, or were, not in a condition that they needed to be taken away immediately. But that was somehow beneath them, they were the ones to decide.

T: Mm. [pause] But you had no one with you at the time when this happened [

I:] no

T: but you were

I: All alone.

T: Mm. Then, did they take the dogs with them?
I: Then they took the dogs with them, by police escort, and, and then . . . ehm.
A car then that came and took them to a dog shelter.

(Interview dog owner)

Within the context of animal ownership, one could assume that the owner would be entitled to the epistemic stance (Potter 2010; Heritage 2012) of knowing about his or her dogs. The dog owner above tried to complain about the decision to take the dogs away. Later in the process, he also challenged the visual cues that the inspectors recorded in the report. For example, it was stated that most of the dogs were more or less permanently confined to crates, and that some of them had tooth injuries from trying to get out. The owner, on the other hand, challenges the amount of reported teeth injuries and other visual cues such as the condition of the dogs' fur. He also gives other explanations, stating that the particular breed of dog has weak teeth for genetic reasons and that the dogs were just temporarily in the crates while he was away from the house. However, the appeals court did not accept his complaints and the owner was prohibited from housing dogs in the future (interview dog owner). In this situation, the owner is not found credible, and consequently, his "believability" (Iversen 2013), based on the recorded sense of the place, is found to be low. Thus, sensuous epistemology is unequally distributed, and the epistemic authority is for many reasons, unsurprisingly, located to those in official positions: they are professionals with high credibility, the animals were found in a miserable situation and recorded as such, and the law was violated.

Recording numbers and/or individuals

In news reports on animal hoarding, numbers seem to be an important ordering device. For example, one local paper wrote, "More than 50 cats lived here in misery" (*Uppsalatidningen* 2011). Animal numbers—also in the social and behavioral science accounts—add to the construction of deviance by the absurdity of the case, for example "Barbara Eriksson and her 552 dogs" (Arluke and Killeen 2009) and thus contribute to the stigmatization.

In the context of detecting animal hoarding, the "number fetish" is not so prevalent. When dealing with potential animal neglect it is important, although not always possible, to take records of the conditions of each and every animal.

Caesar, male
Nils, longhair, grey with white feet, male, tangled fur
Freddy, black and white, shorthaired male [. . .]
Ronja, short haired, female, runny eyes
Sammy, longhaired, male, faeces in the fur on the back, along with tangled fur.

(Inspection report, County Administrative Board X 2011)

The animal welfare officers thus record data, mostly based on visual cues detected by themselves, and sometimes also by an attending veterinarian. It is of course of importance to specifically record cues that point to neglect. One person who got an unannounced call from the County Administrative Board described the situation as such, when asked about the officers' appearance:

I: Yes, right, protective overalls and [
T:] huh
I: stuff. Yeah, but it, you have to be sort of honest and show them around every-where and . . . and so on. So that it . . . it was just a matter of walking around, with them right, and counting the cats in and . . . telling them their names and they took pictures and so on.

(Interview cat owner)

The person under investigation sometimes, as above, assists by providing the names of and/or other data regarding the animals, and sometimes not. Whether the latter is due to a lack of knowledge about the animals or not wishing to coop-erate is not easy to determine, of course. But what is clear is that recording the individual animal data is a task that may seem easier than it is:

Because then, eh, when we had the 170 cats, we put down 150 because they were far too wild [. . .] but, there was a veterinarian, right, who judged who was to be put to death and who was in a condition to keep on living.

(Interview animal welfare inspector)

Obviously, when dealing with large populations of animals in a stressful and crammed environment, the animals themselves seldom cooperate (being "far too wild"). And sometimes it is the condition and quality of the place, in combination with animal resistance, that limits the recordability of the animals:

The light was very limited, for example in the hall, which meant the cats could hide in many places. Countless numbers of cats of various ages were in the dwelling, roughly at least 15 heads. The older ones hid and ran off, while the younger cats were closer. Four of them were caught and examined more closely. Three of them had severe ear mites and running eyes too. One of the kittens was snotty and coughing.

(Decision on immediate custody,
County Administrative Board X 2008)

"Recordability" is a term that refers to the "efficiency in getting particular informa-tion" (Iversen 2012: 694). In an institutionalized situation where inspectors strug-gle between the requirements of filling out a detailed standardized protocol and the reality of a rather messy place, the recordability may be challenged. Sometimes the outcome is almost absurd, in terms of the objectivized discourse used:

Roughly 190 cats are confined in 4 groups in a total area of about 45 square meters. A storage space is not allowed to house more than 15 adult or

20 growing cats. All cats should have simultaneous access to a space where they can lie down [*liggplats*]. Space intended for permanent cat housing should be at least 6 square meters. The area should be at least 2 square meters per cat. Accordingly, there was a heavy overload in the 4 rooms.

(Inspection report, County Administrative Board X 2010)

Numbers of animals are thus important in many ways. First, they give cues as to the state of the place. Many animals in a small apartment or house indicate that there may be a welfare problem. Crowding as a spatial quality of a collective in a certain social and physical setting seems to be an important feature when evaluating human/animal relations, especially in the city. Second, numbers may be important in terms of welfare priorities; having scarce resources, animal welfare officers may have to give less priority to inspections that involve few animals. "Many" is of course a contextual construction, and thus how the feeling of the place is recorded is of great importance. Numerical accounts symbolize unchanging truths; they have an aura of absolute finiteness to them (Bloor 1991 [1976]: 85). However, in this context, numbers play quite a contingent role. Numbering is, thus, in this construction of deviance, an indexical affair (Garfinkel 1967). Contrary to the inherent exactness of absolute figures, the feeling, quality, and size of the particular place play a crucial role in the epistemic process.

Smelling odor, seeing dirt, hearing noise

As indicated above, inspectors make sense of places through their perceptions, their full topology of senses (Serres 2008). One of the inspection protocols, reporting from a visit to a cat owner's home, states:

The police man opened a door that was located in the hall, behind the kitchen door. The door was locked, but could be unlocked. Behind the door was a narrow staircase leading to the basement. A large number of cats were noted on the stairs. The County Administrative Board counted about 30 adult cats. It smelled very strongly, stronger the further down in the space you came. It was a stinging smell of ammoniac that affected your airways. Down in the basement 10 empty bowls were noted. There was neither water nor food in the space. There were 4 cat trays of normal size. The cat trays were full of feces and urine. There was plenty of feces on the stairs and on the floor. There were a lot of flies in the space. There were no windows in the staircases or in the basement.

(Inspection report County Administrative Board X 2010)

Perceptions and descriptions of odors are complex social phenomena. First, odors are embedded in matrixes of power, where typically people of color, the working class, and women in general, to name a few, are groups that have historically been considered "smelly," while white, middle-class men view themselves as scentless. Second, odors are distributed differently across space, and in the urban environment, smelling in private and public domains are perceived differently (Classen, Howes, and Synnott 1994: 161ff.). Third, smells and "odor pollution" are more difficult to measure than, for example, noise (Classen et al. 1994: 170). However, there are ways, and the ammonia level is one such objective scale that is used in this context.

> Inside the front door, a heavy stench hit the inspectors. The floors in the hall, kitchen and den on the lower floor, were soaked with fresh and dried cat excrement. On the upper floor, there was a several-meter-long bank with cat poop and newsprint. In the hall, the ammonia concentration was measured at 10 ppm using a glass tube meter, and 26 ppm on the upper floor.
> (Decision on immediate custody, County Administrative Board X 2008)

Odor pollution is often the reason why neighbors complain to the authorities. When interviewing one person who had been subjected to complaints filed by neighbors due to the smell and noise from her many cats, and who lately received several visits from the animal welfare officers, I asked what she thought was the problem:

> *T:* Well, according to the protocol in, the, the biggest problem was that they thought it was dirty . . . that it smelled bad. What did you think, kind of, what do you say about this [
> *I:*] I agree
> *T:* what, what was the biggest problem, according to you?
> *I:* I had become so neglectful right, with cleaning, in particular and . . . eh, I had well, come home at three-four at night from here [workplace], so I was asleep, and so I hadn't had time to clean the litter trays when they arrived. So they were quite full by then, 22 cats have time to . . . but I had like, had 12 trays running, I think, right then. So, there where trays alright.
> (Interview cat owner)

The cat owner above agrees with the account of the situation given by the County Administrative Board, but provides an alternative explanation of the dirt and smell in the house. As she understood it, lack of time, rather than animal neglect, was the circumstance of importance.

According to the animal welfare inspectors who are in charge of deciding whether animals will be taken away from their owners, it is quite common practice to immediately confiscate abused animals on the spot and, depending on their health conditions, to euthanize them or send them to cat or dog shelters. With the help of some creative twisting of the law, in some cases, and depending on the conditions, hoarders are allowed to keep one or two animals. This helps some of the owners, while others quickly return to old patterns. According to the interviewees, in the end some hoarders even feel relieved when the authorities take over, removing the animals and, in this way, letting them regain some control.

In the end, the various visual, auditory, and olfactory cues need to be translated into written accounts according to standardized procedure. This translational practice is interesting in itself, when people transform, through a number of measurements, indexical categories into objective ones (Kumlin 2011). The welfare protocol becomes a "mediator," translating an action from one part of the network to another (Latour 2005), thus a central actor in producing meaning around the phenomenon of excessive pet keeping. Accompanying the written protocols are photographs of the animals in their surroundings, and sometimes also video recordings. These photographs can be interpreted as yet another set of technologies, capturing and translating sensory impressions into objective accounts. However, they could also be understood as ethereal extensions of animal agency, enabling the animals to reach out and contribute to their own preservation (Hayward 2011). It may be speculative, but I would like to speak about the animals in this context as "sentient agents," communicating cues to be picked up by human senses.

Summing up, the recording of sensuous impressions starts at the very beginning of the visit, even outside the dwelling. The smell of excrement, noise from barking dogs or meowing cats, and the sight of dirt together provide cues as to what to expect when entering the premises. Later, the counting and examination of confined animals—along with objectified impressions and technologies of control such as measuring air quality, temperature, and light intensity, counting numbers of litter trays, gathering photographic documentation, and measuring square meters—contribute to the production of epistemic authority and will, in the end, account for the measures taken. In addition, it is important to recognize the sentient agency of animals, as cues are produced within a particular social setting, thus between human and non-human agents. Moreover, the authorities are keen on pointing out that the hoarders are seldom evil, rather they love animals a little too much and thus take on more than they can care for. In this connection, one may naturally suspect that when the behavior is explained in this way, intervening in any radical way may be difficult. If human neglect of animals is said to be a consequence of love gone awry, it is not surprising if inspectors are soft-hearted. However, this does not seem to be the case. Through this attitude, the deviant behavior is somewhat excused, explained in a way that makes sense to everyone involved, and face saving can take place for an otherwise stigmatized person.

Verminizing through crowding

The chapter has attended to the sense-making of cross-species disconcern: explanations of excessive pet keeping and their "sensuous impressions" (Hayward 2010: 581) in social interaction and the mediated production of knowledge. After reading, listening to, and watching the stories of many so-called animal hoarders, what are the common features? First, these people most often refer to some kind of traumatic life crisis as the starting or tipping point that marks the beginning of their taking on animals in numbers: the death of a close friend or loved one, a major illness, loss of a career/job. The common feature is that the animals helped them get through the crisis, providing a different focus, different values, and endless affection. Taking care of these animals requires plenty of time, effort, emotional involvement, and money. Second, at some point, the collecting of animals reaches a threshold, beyond which it becomes a problem, both to themselves, their family/neighbors/friends and to their animals. All the energy and time that are channeled into caring for other animals create social tensions. People start talking, neighbors take action, welfare authorities get involved, and shelters are contacted. This creates a great deal of negative emotions, feelings of being threatened, and sometimes feelings of shame. Thus, the common narrative portrays a process involving certain tipping points, or turnarounds, the entanglement of many actors and affects, in which the positive dimensions are gradually transformed into burdensome and complicated relations and feelings of shame or guilt. Identities also change through the process, the owners become de-humanized or even animalized, while the hoarded animals lose their pet-identities and become feral, even wild. This verminizing process thus illustrates what sociologists have discussed in terms of strangeness, the dialectical process of becoming stigmatized as outsider. Thus feelings such as shame or disgust are not only personal matters, they are also a matter of social ordering (Howes 2005: 10). For example, odor and morality have a long-standing alliance in strategies to purify the Western social body (Classen et al. 1994: 172), including urban spaces (Sennett 1994). Animals play important roles in the governance of hoarding, in that they already often stand for the unruly, impure, and liminal urban stranger. In the hoarding context, they also perform certain feralness, most often refusing to collaborate with authorities.

In Chapter 2, I elaborated on the concept of "trans-species urban politics" in an attempt to understand the purifying ordering of human and animal bodies in urban public space. In this particular case, however, it is the governing of human/animal relations in urban and suburban homes, through the epistemological work of scientists and experts, and the practice of knowledge production and assessment, that is under scrutiny. As demonstrated, it is a rather messy affair, in which actors use available cues and tools to get a grip on and make sense of the situation. The perceived chaos is rationalized in Garfinkel's (1967) terms, through the use of technologies such as protocols, photographs, and instruments, and interpreted in light of legal texts, thus mediating between sensuous impressions and outcome. The governance of animal hoarding relies both on the response-ability and on the recordability of human as well as non-human actors. The sensorial impressions of, or perhaps rather in the place in question, suggest that it is meaningful to use the framework of emplacement (Feld and Basso 1996). "Emplacement" suggests that bodies and sense-making are always taking place; they are performed in certain places, and thus, these places are enacted. "While the paradigm of 'embodiment' implies an integration of mind and body, the emergent paradigm of emplacement suggests the sensuous interrelationship of body–mind–environment" (Howes 2005: 7). As demonstrated in this chapter, places are sensed and are thus involved in methods of knowledge production. But senses are also placed (Feld 2005). The quality of the place in terms of air, light, area, functionality, numbers, and species of bodies—degree of crowding—and perhaps most decisively dirt, directs and informs the trans-species sensuous impressions. I argue that it is important to attend to the multi-sensorial processes taking place in these encounters, rather than the objectified data that guide authorities' notions of the affordances of the place and consequently the actions to be taken (see Pink 2009).

Reconnecting to the overall framework of the book, that of crowding in zoocities, I argue that animal hoarding can be understood as a verminizing phenomenon: a process through which explanatory models, law, and sensuous governance shape and reshape human/animal relations in a particular setting. When the animals become too many, the relationships too odd, the home too filthy, and the owner too self-neglecting, the species order is threatened. The animals "take over," the human "loses control," and starts behaving in "animal like" ways. Is this transgression due to a failure of the anthropological machine? As Tim Ingold has pointed out, human and animal are separated through their different belonging to culture and nature (2011). While animals, whether domesticated or wild, remain in the natural sphere, humans may move in between. Why is that? Because human existence entails a certain element of animality, which poses a threat to the position claimed through civilization and domestication: "Thus, human beings, uniquely among animals, live a split-level existence, half in nature, and half out; they are conceived as both biological *and* cultural beings, organisms with bodies and persons with minds" (Ingold 2011: 4). Consequently, native groups or, in the present case, animal hoarders, are conceived of as living too close to animals, to uphold their humanity against the inner *animality*. But while native groups are so clearly associated with "nature," surely the spatial context of the city should be proof of the humanity? In this chapter, it is the urban and suburban home that is the particular place under investigation. Within these places, various human and animal bodies—such as dogs, cats, human authorities of different kinds, and so-called hoarders—interact and produce knowledge about the situation in relation to cues such as the size of the dwelling, feces, and torn wallpaper. I suggest that the lack of homeability, along with the norm breaking in terms of human/animal relations—too many, too important—contribute to the construction of deviance. As argued in Chapter 3, homeability refers to a continuum of homeliness effects. Torn wallpaper and cat corridors may add to homeliness for cats and their companions, but not for the hegemonic middle-class norms that will instead consider such interiors as non-homely and as indicative of the inhabitants' animality.

I certainly do not advocate animal hoarding or any other cruelty—on the contrary. But I do think that there is a need to look beyond *DSM-V*, and into the messiness of emplaced relationships involved in urban animal hoarding, in order to understand love, neglect and abuse. There are most certainly promises to this kind of crowding. On a personal level, caring for others, en masse, means the positive need to let go of many existential and life history concerns. For the social animal, it means a promise never to be left alone. On another scale, animal hoarding provides a threat to contemporary human/animal norms, and thus provides a normativity leakage. Attendance to the dialectical interconnectedness of meaning and socio-spatial context, which exceeds species boundaries, is thus fruitful when analyzing animal hoarding in particular and, in so doing, questioning the notion that sentient agency is an exclusively human property.

5 Feline femininity

Emplacing cat ladies

Cat ladies are . . . romance-challenged (often career-oriented) women who can't
find a man.

(*Wikipedia* 2013)

Species reeks of race and sex; and where and when species meet, that heritage must be united and better knots of companion species attempted within and
across differences. Loosening the grip of analogies that issue in the collapse of
all man's others into one another, companion species must instead learn to live
intersectionally.

(Haraway 2008a: 18)

If you Google the string "cat lady," you will be deluged by images of
crazy-looking, deviant, and asexual women surrounded by cats. A good proportion of the representations are satire-like—drawings and comic strips—making
fun of these women's dowdy appearance. There are also many commercial goods
available for purchase: crazy cat lady games, plastic figures, start-up kits, even
soap. A recurring theme is that cat ladies are portrayed as lonely and childless,
(hetero)sexually inactive, often middle-aged or older, sometimes represented as
ill. There seems to be something missing in their lives—a hole that living with
cats may superficially fill. However, the term cat lady also connotes sexy, slender,
luscious, and pleasure-seeking women. This image mainly occurs in commercial
ads and films, and movies. In both cases, however—the dowdy and the sexy—
ingrained notions of felinity and femininity intersect to produce a certain culturally intelligible image. In this chapter, I investigate images of women who live
their lives, often a bit on the margin, surrounded by their cats, as I expect these
images to reveal norms for gender as well as human/animal relations. In short, this
chapter explores representations of cat ladies from an intersectional perspective.

Representations are social relations, and as such, as Doreen Massey points
out, they "always have a spatial form and spatial content" (1994: 168). Thus,
social relations, including representations of human and animal bodies, exist only
through, in, and across space. The urban context of the cat lady representations
discussed here causes the analogous "private" space and "feminine" position
to collapse, making them a public matter of ascribing identity to places, pets,

Figure 5.1 Cat woman and Batman (Cartoon: Michael Heath)

and people. In line with the overall framework of this book, I am interested in notions of zoo and humanimal crowd/crowding in the context of cat ladies and their emplaced cultural representations. Because the prevailing narrations of cat ladies are essentially disturbing (the term "crazy cat lady" is often used in English-speaking popular culture), the notion of zoo is apt if we wish to capture the dialectical disordering through humanimal crowds and the ongoing ordering practices based on cultural purity.

As Haraway points out in the quote above, companion species need to learn to live well together. In her view, part of that lesson is to stop thinking about humans and animals as culture and nature, as "us" and "them," and to start considering all relations as intersectional (2008a: 18). This means, among other things, that species, race, and gender are relational rather than analogous: women, for example, are not (treated) *like* animals in pornography and trafficking; similarly, industrial meat production systems, including abattoirs, are not *like* Nazi concentration camps. These and other profoundly troubling phenomena must be analyzed in their own context, otherwise we will contribute to the downgrading of other animals, and lose some essential insights into humanimal becomings in that specificities may be lost. What is at stake is no less than what Simmel would call sociation—the becoming of oneself in the institutional and physical context of others, here extended to animal subjects. Untangling stories of cat ladies by following the threads of animality, humanness, femininity, and felinity, this chapter explores notions of cat ladies using an intersectional approach, revealing norms and expectations regarding proper behavior, also in terms of class, age, and sexuality.

Of animals and gender

To help me answer the questions above, I use a range of contemporary popular cultural productions, such as TV shows, documentaries, and reality programs, which are primarily, but not exclusively, Swedish. In our current, globalized media environment, it is not only difficult but also unfruitful to set national boundaries for cultural influences and representations. However, I have chosen representations of real women and real cats, thereby excluding material with women dressed up as cats (so called "furries") and fiction (e.g. Cat Woman and The Simpsons' "crazy cat lady"). In addition—following up and adding to the analyses made in Chapter 4—I make use of some psychological texts focusing on women living with too many cats in order to tap the scientific discourse on "crazy cat ladies"—that is, how they are explained and represented. The popular cultural and behavioral scientific material has been rounded out with interviews from the project. When speaking about the intersection of gender and cats—materialized through cat ladies—expert authorities, cat shelter workers, and cat owners have something to add to both explanations and representations. I have read all the texts together, again with an ethnomethodological gaze, looking at the ways in which images are produced: what cues, methods, and resources are used, and with what consequences. These images are not free-floating, but historically specific and situated. They are also saturated with norms regarding species, gender, sexuality, age, class, and sociality. However, as the analyses will show, there are also interesting subversive elements inherent in the image of the crazy cat lady.

To explore representations of cat ladies, we must first untangle how we can understand humans' relationship to companion animals in general and women and cats in particular. Donna Haraway has coined the term "companion species," or interconnected species, to describe species that depend on one another's existence (2003). The term does not designate specific species relationships, even though she uses dog/human as the ideal typical relationship. Dogs and humans have co-evolved, and continue to live with and by one another. These relations have an important power aspect, because even if dogs have many advantages—which have revealed themselves throughout history—of being in humans' proximity, they inevitably become abused and exploited. The critical issue is alliances between species, as in this chapter between women and cats, in which the relationship constitutes a kind of mutual, albeit unequal, dependence. Meanwhile, it is humans, as Haraway argues, who are expected to be responsible for the relationship (2003). This is a counter-argument against a more critical view of pets, in which you basically see them as an expression of human supremacy, a product (Tuan 2005).

But pets are subject to both domination/exploitation and love/closeness (Haraway 2003; Redmalm 2013). For Haraway, companion species are primarily an entry into understanding and explaining contemporary social orders, which are about shaking the very idea of species hierarchies and worldly becomings. As with her previous figures (the cyborg, onco-mouse, and the vampire), the companion species is a "material-semiotic" figure, which can be used to explore the boundaries between nature/culture and animal/human, and to demonstrate the arbitrariness of these distinctions and their permeability. In this way, we can also begin to challenge the ideological discourse of "human exceptionalism" (Haraway 2008a), which in many ways is the background to the destruction and exploitation of the environment and the animal world, a fundamental idea within post-humanism (see also Wolfe 2009).

Similarly, I want to use the "cat lady" as a trans-species, material-semiotic figure, an analytical tool to explore the above-mentioned traffic between categories. With the help of this figure, I also want to go beyond species relations and problematize the dimension of gender. The link between gender studies and animal studies, however, is not particularly articulated (Birke 2002). Nina Lykke argues that one reason for this, in addition to the fact that animals themselves cannot make their case in the feminist dialogue, is the sex/gender distinction (Lykke, in Holmberg 2007: 46). Both cases (sex/gender and animal/human) are thought of as dichotomies with watertight boundaries and discrete categories. Feminist researchers' focus on gender has contributed to not only a "black boxing" of biological sex, but also of animals that fall on the same side of the nature/culture boundary (Bryld and Lykke 2001: 33). However, within the human-animal studies field there are many connections to gender and feminist research, in terms of eco-feminism (Emel 1998), feminist ethics (Donovan 2007), and queer and transgender studies (Giffney and Hird 2008). But when I want to understand representations of cat ladies, it is not only animal/human relationships and the species dimensions that are in focus, but other power structures also become relevant. I am primarily thinking about gender regimes and expressions of femininity, but class, age, and sexuality are also present in my encounters with cat ladies. Within gender research, a fundamental idea is that gender is something that is made, rather than something you have: gender is performative (Butler 1990). In practice, gender is made in a variety of ways, partly depending on other performative qualities such as class, ethnicity, and sexuality. Here the concept of intersectionality is helpful, as is the fundamental idea that various power structures work together, "intra-act," in practices and discourses, and thus also transform one another (Lykke 2005: 8). The purpose of using intersectionality as a point of entry is to achieve "an analysis of how the categories construct each other" (Lykke 2005: 8). Lynda Birke and colleagues (Birke et al. 2004) have developed Butler's performativity theory through the notion of "animaling," thus including species. Their terminology is useful to talk about how animal/human relations are made, or animaled, as well as how humans and other animals can be animaled differently through gender, class, and sexuality. Thus, I want to highlight how relational and close interconnection between species is at the intersection of different categories and orderings. To refer back to Haraway's terminology, I argue that species, gender, age, and class meet, and are made visible, in and through the material-semiotic figure of the cat lady. The cat lady is, in other words, both an empirical and analytical figure.

In what follows, I will, after a passage delving into the cat's history, and images of women and cats, explore these images in terms of the lonely, the eccentric, the victim, and the rescuer, and finally discuss them in light of sentimentality as an ethical position, and "feline femininity" as an alternative or resistance femininity.

The cat's place

In Gösta Knutsson's wildly popular novel series about Pelle Svanslös, set in 1950s Uppsala, Sweden, we find many different cat characters: Måns, the homeless, urban alley cat; Murre, the hillbilly cat from Skogstibble; and the house cat Pelle, tame, tailless, and (at least a little) harmless. Pelle's gender is, at least initially, unclear (Knutsson 1939 [2014]). (It has been argued that these characters were based on real-life characters in Knutsson's surroundings, but that is another story.) Similar to Knutsson's gallery of characters, real cats are framed in different ways in the community. The "cat-egories" are characterized and understood in different ways based partly on their relations to humans (homeless, feral, domesticated) and to the places in which they move (house cat, domestic cat, farm cat, city cat) (see Chapter 3). The urban cat's proper place is in the home. But this has not always been the case. The Scandinavian cat has probably been domesticated for about 1,500 years (Broberg 2004). Despite this, we do not know much about its background. Gunnar Broberg, author of *The Cat's History* (2004), highlights how the cat is surrounded by rich symbolism—cheating, deceitful, sensual, frivolous, but also enchanting and associated with dark forces—and the magic lives on throughout history (Broberg 2004: 46). But the cat is also feminine, and the link between the woman and the cat, through motherhood, sensuality, and domestication, was established early on. Moreover, just as the woman is seen as an imperfect man/human, the cat is seen as an imperfect dog/pet. Positive values are also stressed in both the cat and the woman. Both are associated with the home, adeptly guarding over the house and waiting with food for their masters to arrive. The cat is often portrayed as the ideal mother who selflessly cares for her young. In his book, Broberg describes various historically famous Swedish cat ladies. Elisabeth Margareta Hjerta, who died in 1811, "was widely known for her extraordinary love of cats, of which she had a whole menagerie" (2004: 91; author's translation). The "cat madam" in Åsberg is another example. Common to them all is that they are described as weird, odd, and that it is the number of cats they keep and the nature of the relationship that often makes them strange (2004: 92). Cat ladies embody a "feline magnetism," at once untamable, mesmerizing, and mystic in their outsiderness. Interestingly, Broberg argues that cat ladies may have come to be reappraised in late modernity, and that, alongside a feminization of positive values, we may be observing a felinization, a shift toward upgrading the feminine/feline while the male/canine is seen as hopelessly traditional (2004: 351).

The question that arises is what this means, and where the boundaries are drawn for the feline presence in humans' lives. My hope is that the cat lady as a trans-species figure can illustrate the tension between legitimate and illegitimate animaling, and thus say something interesting about the more general cultural context. In the following, I investigate images of cat ladies in popular culture, behavioral science, and interviews with public officials, cat shelter enthusiasts, and so-called cat hoarders.

The lonely

The stereotypical cat lady is a culturally pervasive image. Ever since the medieval fears of and hunts for witches, the woman who communicates with cats has stirred emotions like fear, disgust, and hatred, but also pity and ridicule. In order to understand the pervasiveness of these stories, one needs to explore the relationality of species and gender, of cat/woman through the label cat lady. But other dimensions come into play as well. Demographical studies, although most often not based on randomized data, confirm that a majority of so-called hoarders—people who collect animals—are women, over forty, single and with income and education levels below average (see, e.g., Patronek 2008). When asked about their experiences with people who have many cats, the police officers in my study rehearsed some of these characteristics:

I: I've never experienced that someone really like, eh, from the upper class had, eh . . . lots of cats.
T: no, there is a certain class [
I:] Yes.
T: class pattern [
I:] Yes.
T: in that.
I: Yes, I believe it in fact.
T: And gender pattern.
I: More women. Although we have still had one little old "cat man."
(Interview police officer)

Interestingly, almost all the informants contested the gender-stereotyped notion of the cat lady. In a country where gender equality has been a strong official norm for decades, gender-specific norms are difficult for interviewers to pinpoint, as in the quote below, in an interview with two animal police officers discussing the prevalence and characteristics of cat hoarding in the city:

T: Are there any gender patterns?
I1:   No . . . we've had both older women and older people, most often [
I2:   ] Older, yes.
I1:   older pensioners in fact. You can see that, anyway. With both men and women who have . . . but it has similarly gone [

I:] Mm, mm.
I1: out of control. But [
I2:] But then [inaudible] something like this it's more that they are alone, but they are [
I1:] Yes.
I2: sick also [
I1:] Yes, precisely.
I2: but that they, one is alone, one wants to have . . . have some companionship and so then I guess they go a little overboard.

(Interview police officers)

Obviously, it seems reasonable to try to make sense of a norm-breaking behavior by relying on available interpretative repertoires concerning lonely, ill, and elderly people who want to have some company and someone to care for. Note how the interviewees co-produce the notion of the older person as being alone and wanting some company. A recurring theme is thus that cat ladies (whether male or female) are portrayed as lonely and childless, (hetero)sexually inactive, often middle-aged or older, and are sometimes represented as ill. There would seem to be something missing in their lives, something that living with cats may remedy.

In the Canadian documentary *Cat Ladies* (2008), we get to meet four radically different cat ladies, women who define themselves in relation to their cats, but who nevertheless are "normal" ("I am not a crazy cat lady," as one of them expresses it). This statement, of course, says something essential about the available representations and stereotypes. The film's website contains the following text:

> It's not the number of cats that defines someone as a "cat lady," but rather their attachment, or non-attachment, to human beings. They create a world with their cats in which they are accepted and in control—a world where they ultimately have value.
>
> (*Cat Ladies* 2008)

In other words, the film-makers show how women who live alone get a sense of value in relation to their cats. I will problematize the idea about control later; until then we can conclude that cat ladies are defined in terms of lack of human social attachment. Part of the film is about Margot, a middle-aged woman who works full days and then comes home to her three cats. Her homely city apartment is decorated with ornamental treasures, cat portraits, and show rosettes. Her identity is portrayed so as to imbue it with her relationship with her cats—both present and past. In one scene, Margot tells of her sorrow over dead companions. The cats in turn are portrayed as entirely harmonious and well groomed; they lie in Margot's lap and purr, rub up against her leg, and eat appetizing foods with long, pleasurable licks. When Margot comes home from work, she sits on the sofa and talks with her cats, who all have individual names: Bongo, Fritz, and Little One. Naming is a symbolic act that transforms and elevates an individual—in this case the cat—to a person (Sanders 2003). The cats are truly persons—Margot's friends and confidants—or so it is presented, up to a certain point. Then she begins to talk about her lack of human friends and family, swallowing hard and saying with sadness in her voice, "a lot of people don't know, that I am as lonely as I am" (*Cat Ladies* 2008). Here Margot suddenly defines her loneliness in relation to humans, and cats are presented as inadequate substitutes.

In contrast to Margot's harmonious home and happy cats, later in *Cat Ladies* we become acquainted with Diane and her 132 cats. First the viewer is taken around a chaotic house, where we follow the counting of cats. Most are locked up in cages, but quite a few of them are running around, up and down behind furniture, crawling under beds and into drawers, jumping around. The cats stroll about, pulling their ears back, wagging their tails. They throw themselves onto their food and gobble it down, while Diane is constantly on the move, attending to the cats and cleaning. She changes litter boxes and cleans floors, opens cat food jars and gives the cats medicine. Diane is more or less forced to get up several times each night and clean, even if she has help from a volunteer who comes in at night. We hear that Diane previously had a good job; she wore suits to work and often traveled away on business trips. When she was laid off, she began taking care of homeless and abandoned cats. Here one can imagine a case of downward class mobility, where the suit marks a life that has been, something that has been lost given the presence of the cats. But the suit is also described in terms of femininity, when it is contrasted with her current unglamorous appearance ("Look at me now!"). The respectability of a middle-class appearance has been lost. Now the problem has grown to the point where it has gotten out of hand, and in an emotional scene Diane says that she must stop helping the cats and start thinking of herself. She wants to travel and visit her relatives, but is unable to do so because of the cats. A situation is presented in which Diane has lost control of her life, and the cats now control her. This is a recurring theme in many stories about so-called cat hoarders—people who have too many cats and thus cannot take care of the animals—and themselves—in the way society expects them to (see Chapter 4).

Sociologist Clinton Sanders (1999) talks about how, in his case, people and dogs are constructed and assessed together through the attribution of interspecies identity. Thus, dogs' behavior impacts how we perceive humans' behavior, and vice versa. Taking the idea a step further, Harlan Weaver starts from a trans/ pit-bull perspective and discusses the notion of "becoming in kind," as a process through which the co-shaping of identities includes complex webs of species, gender, sexuality, class, and race:

> In moments when my appearance has been at its most liminal, when I have felt vulnerable as a visibly transgender person, she has ensured my safety. Concurrently, my whiteness, queer identity, and middleclass status encourage other humans to read Haley as less threatening; in my presence, she is perceived as less dangerous. Each of us shapes who the other is. This enmeshment of our identities exemplifies what I term "becoming in kind."
>
> (Weaver 2013: 689)

In the case of Margot and Diane above, it is evident that the film production exploits such trans-species identity markers. The stressed Diane rushes around just like her cats and although they likely don't have much choice as they are trapped in the house, even Diane is portrayed as a victim, held down by the overwhelming amount of cats. Gender and class interact in this animaling process, so that Diane is made into the helpless woman in relation to the cats who take over, and she is said to have ended up in this situation after losing her professional status. Margot, in contrast, is portrayed as an inadequate woman when she "mothers" cats instead of children of her own. Moreover, the place called home looks very different in these two stories. On the one hand, pictures of Margot's feminine apartment, full of personal decorations, pillows, and framed photographs, intersect with her cat mothering identity, while Diane's chaotic house, portrayed as totally dominated by the lives and bodies of cats, produces a troublesome identity: a woman without a home, but with a house taken over by cats.

If the women lose their human-ness due to their closeness to the cats and the lack of human relationships, the cats lose their pet-ness. Hissing cats in cages are at odds with our idea of the domesticated, docile pussy-cat. But even the human-ized, babied cat, who shares her "mother's" sandwich, breaks with the ideal image of the pet. The examples clearly show that pet categories as well as human ones are contingent positions—their boundaries can be moved and renegotiated. They are also made in complex dialectical interactions with one another, in a relational dance of becoming with.

With help of the cat ladies, and congruent with this line of reasoning, I would like to problematize the prevailing sociological notion of sociality. Instead of see-ing it as a zero-sum game in which humans' relations to other animals are expected to come at the expense of "real" human relationships, it seems more reasonable to analyze and evaluate all types of social relationship in a more symmetrical fash-ion (Serpell 1986). In accordance with Haraway, we can realize that "to be one is always to become with many" (2008a: 4), a subject formation that takes place in the context of multi-species relations. This of course requires an extension of what is usually meant by "social," an enlargement that cat ladies may facilitate.

The eccentric

We will now move from Canada to a small town in the Swedish region of Dalarna, and meet Mia through the Swedish TV series *Luxury Trap* (*Lyxfällan* TV3, pro-duced from 2006 with 15 seasons so far). The TV show features people who are regarded as careless with their finances, arguably with the aim of giving par-ticipants help and advice. Two financial experts, Magnus and Patrick, come to the participant's home, and those who follow the show know that, even if the two coaches' advice deals primarily with financial solutions, the participants also receive quite a bit of social advice along the way. The episode in question features Mia, who has her own cleaning firm, and who, according to the program, "prior-itizes birds, cats, and teddy bears above the debts she owes" (*Lyxfällan* 2011). She lives in a small detached house with a garden, and her income from the business is spent mainly on animals; she rescues and care for homeless cats through a cat shelter, and feeds wild deer and birds on her property. Thousands of Swedish kronor are said to go to this every month, something that seems to make the hosts nearly speechless. In earlier episodes we have met people who buy luxury cars, home electronics, and home appliances on credit, offenses that were not as bewildering. Yet Mia's interest in the welfare of animals, and her eccentricities when it comes to her collected treasures—teddy bears and lifestyle magazines—seem to be incomprehensible to the suit-clad men.

Here, we can sense a class and gender conflict. Collecting teddy bears and rescuing cats instead of paying corporate taxes seems highly irrational. Helping needy individuals of other species is at odds with the more egoistically oriented consumerism that otherwise tends to be a source of problems. The coaches thus stand for the rational, while Mia's concern for animals and home decoration trends, similar to Margo's lifestyle, presented above, mystifies the male visitors. The cats are curiously absent from the program, represented only as excess, not as individuals or actors. At the same time, we could argue that they still affect the story as a kind of present absence, a constitutive silence, one might say. One of reality TV's main functions is to educate participants and viewers about middle-class norms and values (Skeggs and Woods 2011). This is often done in ways that exude class contempt—the middle class's contempt for workers who are not sufficiently subdued, not tasteful or "respectable" enough (Skeggs 2000). Irrational excess in home decoration—breaking with the Scandinavian interior design aesthetics of light and minimalistic—combined with a lack of interest in contemporary consumption patterns, as represented by Mia, is viewed with disapproval in these educational programs. Moreover, helping out cats in need is portrayed as feminine, and rural emotionality contrasted against the masculine urban rationality—in suits.

A similar story unfolds as we move from Dalarna to Stockholm. In the book *Pictures of Stockholm* (*Stockholmsbilder*), ethnologist Anders Björklund (1994) describes the southern suburb of Farsta, built in the mid-1950s in a functionalist spirit, by painting a portrait—in words—of Inger. She is a middle-aged woman living in a homely two-room apartment near the center of the suburb, with her partner Erik and five cats. Björklund calls Inger the "Catmother," a nickname that in Swedish traditionally signifies a cat lady. She takes care of homeless cats from nearby shelters, preferring especially those who are difficult to rehome, such as older ones. Inger cares for her cats and mourns them when they die, and much to her neighbors' dismay, she buries the deceased cats in the communal garden. She also despises dog owners who neglect to pick up after their pets, and keeps track of the deer populating the neighborhood. According to the story, Inger's interest in animals began as a reaction to her father's mean behavior toward animals and, reading between the lines, his treatment of her. But Björklund, the author, is fascinated by the apparent paradox in her engagement with animals, given that the Catmother does not seem to care much for humans—especially the new Swedes in Farsta such as the pizza baker and other immigrants, whom Inger views without sentimentality or pity. Inger is represented as eccentric; her love of animals makes her different and this love seems to be in need of an explanation: early experiences of abuse, childlessness, sentimentality, and femininity. And her supposed lack of care for human outsiders becomes provocative. Björklund writes:

> How on earth can I put together a picture of Inger? A home-loving lady of 55 years. So warm and sweet to cats and dogs, so short and stingy with the difficult-to-control unknowns. And then this abundant original talent for decor that puts such life into the otherwise mundane apartment. And the equally abundant narration style, which animates every handrail, Gabonese door, curb and clump of grass. I must know more about her, about her life!
>
> (Björklund 1994: 162; author's translation)

Björklund's essay is intriguing. It is full of heart and colorful descriptions—sometimes painting the picture positively, sometimes more negatively with terms such as "stingy" and "mundane." But the story makes me wonder. What role does Inger play in the narration of a suburb such as Farsta? What is it about the Catmother that makes Farsta "territorialized" (Shields 1991: 95) in certain ways and not in others? Built in the functionalist era and into the vision of the "green line" underground, connecting a number of new suburbs on the southwest side of the city center, Farsta is a typical case of modern Swedish urbanity. It is a rather spacious suburb centered on a prefabricated concrete commercial and office building center, surrounded by mainly three-story apartment buildings, playgrounds and green areas, surrounded in turn by small detached houses on the outskirts. Typically, there is not much space for cars. The idea behind the planning was that modern working- and middle-class people would thrive because their whole life world would be concentrated in the neighborhood. They would work, live, and socialize within Farsta. As a step in decentralizing urban life, satellites such as Farsta and Vällingby were products of the new modern vision in town planning, in Swedish called the "ABC town" ("Arbete, Bostad, Centrum," meaning "work, housing, community," see Figure 5.2).

Today, this decentralized unit represents a closed and rationalist image, a place for those who belong and not for others. But, as Rob Shields points out, place images are historically shaped through contested social spatiation, a dialectical process of form and action, space and subjectivity (1991: 18). In this context, the story of Inger the Catmother functions so as to connect the modernist idea with prevailing conservatism and xenophobia. When representing Farsta through Inger, her generosity toward cats and other animals, which is contrasted to her suspicious attitude toward new Swedes, intersects with historically produced place images of Farsta as prefabricated and closed.

I would like to contrast this image with another one in which the cat lady is turned upside down in that it features an old cat man. Donna Haraway's introductory quote on intersectionality is perfectly illustrated by the film *Let the Right One In* (in Swedish "Låt den rätte komma in," 2008; US remake *Let Me In*, 2010). The Swedish version was directed by Tomas Alfredsson and based on the novel with the same name written by Johan Aijvide Lindqvist (2004). It is essentially a

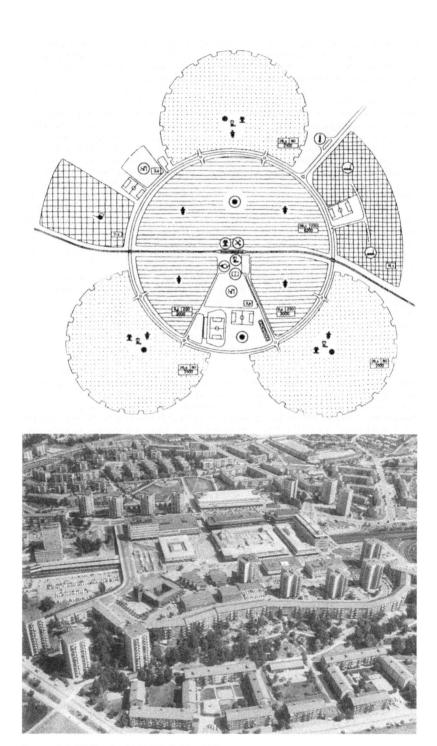

Figure 5.2 Vällingby 1960/ABC City 1952

narrative of vulnerability, marginality, and friendships across boundaries of kind and kin, unfolding and taking place in a dark, winter cold, snowy and highly anonymous neighborhood. The main character Oscar, who is around twelve years old, and lives in a typical scaled 1950s suburb of Stockholm—it could well have been Farsta—is severely bullied and called "the Pig" by his classmates. He is lonely and alienated, until he finds a new friend, a girl called Eli. But Eli is a strange creature: she is utterly pale and untouched by the cold, snowy surroundings, is only out at night, and neither eats nor drinks—except for blood. However, these strange-nesses do not seem to bother Oscar too much. He is energized by the odd friendship and as the relationship develops, he starts to fight back when provoked by his enemies. As a parallel, a narrative of a group of other odd existences is tied in. These middle-aged characters frequent the anything-but-posh tavern Sun Palace: Lacke, Gösta, Jocke, and Virginia. They drink and smoke, tell stories, and socialize. When one of the men, Jocke, becomes a victim of the bloodthirsty Eli, a murder witnessed by the cat man, Gösta, a quiet search for revenge begins. The film is often portrayed as a romantic and violent vampire story. However, it is more than that. It is foremost about alienation, friendship, and revolt, about "underdogs" who through companion species relations try to understand and navigate in a deeply unfair and heartless world. As such, it illustrates the intersectional relations of interest in this chapter:

> The figure is the vampire: the one who pollutes lineages of the wedding night: the one who effects category transformations by illegitimate passages of substance: the one who drinks and infuses blood in a paradigmatic act of infecting whatever poses as pure: the one that eschews sun worship and does its work at night: the one who is undead, unnatural, and perversely incorruptible. . . . A figure that both promises and threatens racial and sexual mixing, the vampire feeds off the normalized human, and the monster finds such contaminated food to be nutritious. The vampire also insists on the nightmare of racial violence behind the fantasy of purity in the rituals of kinship.
>
> (Haraway 1997: 214)

The "underdog" has many overlapping similarities with the vampire: It is composed of hierarchically distributed values—"under"—and disturbs the species order through the noun "dog." It promises some post-human, feminist-influenced connections that in a flash can shed light on the situation of the film's characters. As a metaphor, "underdog" signifies "one who is in a state of inferiority or subjection" (*Oxford English Dictionary* 2012), while *Wiktionary* (2014) also recalls "a competitor thought unlikely to win." In popular culture, the figure is used to narrate well-deserved revenge. Interestingly, "underdog" seems to have originated from dog fighting, referring to the dog that lost, as opposed to the "top dog."

Returning to cats and *Let the Right One In*, the discarded Virginia has been attacked and bitten one night by Eli the vampire girl, but has escaped the incident alive. Virginia is now transfected and thirsty for blood, and stands outside Gösta's apartment house, looking up. Inside, Lacke sits on the sofa and babbles on about nothing in particular. Suddenly, one of the numerous cats raises its hackles and frenetically starts hissing. It is Virginia's arrival that the cat experiences, and her entering the apartment—a passage that alludes to the title of the film—creates chaos among the feline inhabitants. They scream, meow, hiss, and raise their hackles—with their eyes wide open, tails erect, teeth exposed, and claws extended, they encounter Virginia's staggering steps, her clueless and crumpled animal-like appearance. Her boyfriend Lacke anxiously rushes to meet her, while Gösta retires to the balcony with one of his cats: he too can sense the cats' anxiety. Defenseless, Virginia is attacked by biting cats—one, two, three, ten—who with common force, acting as one body, push her out of the hall and down the stairs. The cats have revealed what no one else has seen, and ejected what should not have been allowed to enter. The scene highlights Harawayian ideas concerning the intra-action of different power dimensions. Species, class, place, and gender meet and intermingle through the old cat man; he is also an underdog and a male version of a cat lady. In the film, Gösta's home is full of signs of excessive cat keeping: a pile of feces on the floor, cats on the table, the sound of innumerable feline individuals, and, one may suspect, an unpleasant smell. Citing well-known stories of cat ladies, he is portrayed as middle-aged, lonely, tragic, odd. Likewise, Gösta can be viewed as a queer character who challenges the male heterosexual and virile norm. Through the alliances built across different species boundaries, he as well as the cats disobey hegemonic rules concerning human/animal relations. He becomes more feline than human, since he prioritizes and communicates with his cats; he too can sense that there is something profoundly wrong with Virginia's appearance. Through a similar logic, the cats are also portrayed as possessing more human-ness than the human actors do: the species relations are uncomfortably turned upside down when animality is valued higher than human-ness. While the humans are unable to read Virginia as a threat, the cats take charge and cast themselves upon her, literally throwing her out. As species is about relationality, and human-ness is not an essence, but a contingent and ascribed relational quality, it is interesting to note that it is the societal outsiders—the feline underdogs who have no names but nonetheless possess courage and agency—who are primarily portrayed as human.

The qualities rationality and control are often seen as specifically human. But, what is more, they are seen and coded as masculine. When the cat ladies are portrayed and emplaced, as above, as eccentric, irrational and out of control, it is not only their human status that is challenged. As a consequence of a dichotomous and hierarchical gender order, femininity—whether performed by a female or a male body—is also consolidated as a shortcoming (Hirdman 2001). As will be shown in the next section, this presumed loss of control also impacts how cats are produced.

The victim

Cat ladies are mentioned in the scientific literature as mentally abnormal or ill, and often as victims of early abuse. Sociologist Arnold Arluke, who is one of few social scientists to have written about so-called animal hoarders (Arluke et al.

2002; Arluke 2006), tells about Barbara Eriksson and her 512 dogs (Arluke and Killeen 2009). Eriksson explains in the book about her childhood experiences as a poor girl in a loveless home with a grandfather who subjected her to sexual abuse, and how she received comfort from the family dogs (Arluke and Killeen 2009). Similar stories are repeated in the available psychological studies of animal hoarders, although it should be noted that such studies are very few. It is in fact remarkable how few studies there are in which the individuals themselves are heard. In one such study, we hear from Pamela, a middle-aged, formerly successful filmmaker, who suffered from mental illness and began to accumulate cats (Frost and Steketee 2010: 119). A severe complication is that Pamela's psychiatrist is also a cat hoarder with more than 600 cats, and this pulls Pamela down further. When Pamela's situation was at its worst, she had 200 cats: "Looking back on it, Pamela saw that many of her cats were suffering. 'I was careless with them. I did the same thing to the animals that my mother did with me,' she said" (2010: 127).

The psychologists Frost and Steketee, who for many years conducted research on so-called compulsive hoarding, believe that, based on the few studies that exist and their own interviews, there are certain common characteristics among people who hoard animals. Besides being mostly women, middle-aged or beyond, single, divorced, or widowed (cf. Patronek 1999; Reinisch 2008), Frost and Steketee argue that the women spoke of a "special feeling for the animals," a sense of belonging and understanding that is often difficult to articulate (Frost and Steketee 2010: 129). Animals are seen as more worthy and purer than humans, and love for the animals as "beyond love." One of the interviewees in Frost and Steketee's study is said to have "sheepishly" admitted that she cared more about her dogs than about her husband and her children (2010: 129). Their home had been "handed over to the animals," who "appear to have greater privileges than the human inhabitants" (2010: 129). Once again, you can almost hear the dismay, the surprise that animals are more important than people. The women seem to take on cats to fill an emotional void or to find comfort after a traumatic background, and in relation to cats, the woman restores some control that was earlier taken away. An interesting tension is that, although the overwhelming love of cats is understood as a means to regain control in psychic life, it also contributes to the ultimate loss of control in terms of place, of home. The woman, as a victim of early abuse, thus risks being abused by the cats that are taking over her home.

As a telling example, consider the American reality show *Hoarders* (AETV 2011) in which animal hoarders are portrayed as though they have lost control and the cats have "taken over the house." On the website, a section describes Stacey and her group of animals: "Stacey has 2 days to get rid of her 47 dogs and cats that have taken over her home. They have torn everything to pieces and forced her 14-year-old daughter to move into her brother's home" (AETV 2011). Here it is the pets that have taken over and forced the daughter to move, a statement that of course becomes absurd when the program shows the animals in their small cages. In behavioral scientific texts, as well as in reality shows, it is the cats that take control and force the women, and eventually family members, to step back. However, we need to critically ask: Who are the victims here, and who is the perpetrator?

It is as though the drama requires a clear perpetrator, and it cannot be the woman. She cannot, as it were, be both a victim of serious mental illness and abuse, and an abuser, an animal tormentor. Passive cruelty is a term often used, and one that may allow such sidestepping of responsibility (Arluke 2006). According to Arnold Arluke, who also writes about the sensational media images of animal hoarders, it is seldom the hoarder who is portrayed as the perpetrator (Arluke et al. 2002). Instead, signs of filth and misery are highlighted in sensational pictures and statements. Swedish reporting is similar, and the love of animals, self-sacrifice, and loneliness contribute to the image of women as victims. But what about the cats? Is it ancient images of the wily, deceitful, and self-indulgent cat that sneak in and prevent the cats from filling the victim role? As mentioned above, there is a history underlying the idea of the cat as independent, and today there is often talk about being "cat owned"—as one of my cat ladies put it in an interview—rather than being a cat owner. Again, it seems like "cat" and "woman" intra-act, they construct each other relationally: if the woman is the victim the cat becomes the abuser, while if the cat instead is portrayed as a victim of neglect, the woman's behavior is explained in gender-loaded terms. She loves too much or seeks solace for past abuses and lack of love.

The rescuer

One form of cat lady that I have come in contact with, are those who work with abused or abandoned cats in shelters or foster homes. It is well known that the vast majority of people working with vulnerable animals, internationally, are women (Markovits and Queen 2009). But how gender is performed in this context, and in relation to cats and place, is less explored. When and where is caring for animals accepted, and when and where is it not? What are the fine lines that allow some people to devote all their spare time to cats and dogs through the charity work of shelters, but disqualify (the same) people from doing so in their own homes? What is the institutionalized and emplaced context?

In reality, the boundary between "cat hoarder" and "cat rescuer" is subtle, pointing at a complication regarding the moral economy of feline homelessness (see Chapter 3). It is well known that animal hoarders not only help cat shelters with overpopulation, but they also often care for cats that others have left at the shelters. Neighbors and others know that the person cannot say no, and take advantage of this to get rid of their unwanted cats (Arluke and Killeen 2009; Frost and Steketee 2010: 124). Moreover, both hoarders and rescuers take care of feral cats. In the following quote, I ask one of the shelter workers how she started to care for homeless cats:

I: But, eh it began simply with . . . eh I rode my bike up to . . . and it was spring and late and I looked into a grass grove, a big one like this quite big open grove, and there were a bunch of cats, I think it was seven, eight. But I thought, eh, why are they there huh? [. . .] So, eh, that was probably how it started, it opened my eyes, and then so after that, I thought I saw cats everywhere. And then I thought that one must do something about it, as much as you can. So I started like that on a small scale and advertised in the local newspaper for anyone who was interested and . . . in joining in, kind of, to create a nonprofit organization and then it kind of grew stronger and . . . and so, it went very eh . . . it took quite a long time eh . . . before one could move along a bit and so we started with foster homes and uh . . . yeah and such [

T:] Mm, so it, so it's like just a nonprofit organization?

I: It's just a nonprofit organization. We get no, no money from the municipality, nothing from the municipality. So, we have a small private animal shelter, in the home of one of the board members.

T: Yeah.

I: And she has been involved in this, well, her whole life. She is just over 50. So she has a shelter there with, between 50 and 80 cats.

(Interview cat shelter worker)

During the interview at the cat shelter, 20 or so cats roamed freely and sometimes abruptly interfered with the interview process: meowing for attention, jumping into the lap of the interviewee, or walking across my notes. Looking at the quote above, there is obviously a fine line between officially appropriate care for cats within the legal and institutional frames of a non-profit NGO, and an excessive and eccentric overload of cats in one's private home. The chair of the shelter is depicted as dedicated, rather than as having lost control. Interestingly, cat shelter organizations are hybrid in many ways (see Chapter 3), also in spatial terms. Typically, a shelter can be run from a common building, which may resemble anything from a vet hospital or laboratory to a cat home with individual or co-housed large cages that are open all day long. Most often, the cat shelter involves an elaborate network of foster homes, each caring more or less temporarily for one or more cats. Or, as in the case above, the shelter is placed in the home of the members. One of the women I interviewed in her capacity of cat hoarding (see Chapter 4) had also worked for a long time at a cat shelter. She said that she took on more and more cats in her role as a foster caretaker, until there were eventually so many that "it was simply too much, I couldn't manage it" (interview, cat owner). Another interviewee, who works for a cat shelter with an outspoken animal rights agenda, said she could completely understand how people could end up in a hoarding situation:

T: This thing with hoarders then, animal hoarders or cat ladies as you call it, have you come in contact with anything? [. . .]

I: Um . . . and I understand that people can do drugs, and I understand that people, like, you know, it's hard to say no like that. Because then you end up there and then suddenly you're left with 50 cats and, and the situation is totally unmanageable. And your principles are that you still have to try and to stop buying food for yourself, you buy only food for the cats and I can really see it happening to me.

T: Have you done that?

I: Yes just a little, like when, when, we probably had at most 35 cats, I think, and as I said then, I felt that the situation was like . . . this is not good for anyone and we need to start cutting down and such.

(Interview cat shelter worker)

It feels nice to help out, so nice that the interviewee talks about it in terms of "drugs." Helping the abused animal to a new and better life is itself a powerful motivator that may become addictive. It is interesting how this story differs from and challenges the authorities' "loss of control" image and the behavioral scientist's "compulsive behavior" image, and instead describes the needs of individuals. One of the cat hoarders I interviewed wrote in an email: "You asked what drives me to spend so much energy, time and money on animals. Animals are pretty helpless in the face of human violence. Nature is shrinking more and more because mankind is taking more and more space." In the interview with her, she criticized not only the culture of consumption, but also used an ecological discourse that diminishes humans. Instead, other values were mentioned as important, such as the environment and caring (both for animals and humans).

The rescuer is interesting. This cat lady prioritizes the needs of others, sometimes at the expense of her own needs. She works without pay, and relentlessly collects money and rehomes cats, but within the confines of the cat organization she also teaches others to "speak cat"—to understand and communicate with feline companions and thereby prevent rehoming. One interesting complication concerns feral cats, an urban "cat-egory" that is sometimes handled by the non-profit cat shelters (Holmberg 2014a; see Chapter 3). These cats are rarely receptive to the care, and they sometimes refuse to become tamed. The animaling intermingles with a certain kind of female care rationality, a kind of care that might sometimes go overboard. But the care needs to be received; the ideal domestic cat is, despite its independence, also seen as dependent. A cat that does not want our closeness—for example, a feral cat—lacks in homeability and loses her domestic animal-ness.

Feline femininity: power and resistance

In this chapter, I have described several different common imaginaries of cat ladies and their cats. I assumed that, just as we understand gender and class as socially constructed, we also make animal/human relationships and places through practices of animaling (Philo and Wilbert 2000; Birke et al. 2004). But I also believe that these different performativities intra-act in intricate ways. Sometimes, as in the case of class, cat ladies seem to be animaled differently, while the gender dimension remains static. She is understood as sad—alone and withered—while the cat seems to lose some of its pet-ness. In the behavioral science and agency representations, women are described through the notions of loss of control and victimhood. It would seem to be difficult to see someone who is a victim as being in charge. Instead, the cats are produced as if they were in control. How are we to understand the complex and simultaneously quite unambiguous representations of cat ladies? Are they to be understood as stereotypical representations of the sentimental woman who cares more about cats than about the things one (a woman) should value in life: a clean and orderly home, children and a husband, meaningful work, valuable feminine things? Or do the representations say something more? I would like to suggest that they tell of contemporary connections between emotionality, felinity, and femininity, and I wish to develop the analysis by discussing this coupling in the past and the present. I will examine whether it is possible to understand this connection and love of animals as an ethical position, and resistance of a femininity that questions consumer culture and its callous striving for material happiness. Can one of the promises of trans-species crowds be that they enable one to let go of certain "musts" in life?

Historically, there is a strong connection between the care of animals and women, as well as between the fight for animal welfare and rights and women's struggles for liberation (McAllister Groves 1996; Gaarder 2011). Both battles are united in a fight for the powerless, the defenseless (Dirke 2007). Feminist animal ethics speaks of care and love as the fundamental values that should be used to valorize animals in society (Cuomo and Gruen 1998; Donovan 2007). But the traditional animal ethics and philosophies assert, on the contrary, that emotions can cloud rational, logical science. Peter Singer tells a story about a woman he met who talked about her love for her pets while she happily munched on her ham sandwich. In Singer's interpretation, this demonstrates an irrational sentimentality, something from which proponents of animal ethics must distance themselves (cited in Donovan 2007: 58). Rosi Braidotti and Donna Haraway have both criticized these animal rights discourses. Like others, Braidotti (2008: 106) reminds us that the liberal, humanist perspective that animal rights discourses ultimately build on is saturated with masculine ideals and norms, and is hopelessly anthropocentric (see also Wolfe 2003; Acampora 2006; Donovan 2007). In addition, Haraway argues that animal rights discourses are the siblings of right-to-life arguments (2008a: 297).

The questions that arise in relation to Singer as well as the stories of cat ladies are: Is it necessarily sentimental to talk about and show emotions? And what exactly is wrong with sentimentality? Surely you can understand the love of animals as an ethical position (Rudy 2011), a concern for the needy in a context where both femininity and feline-ness are disparaged? In her book *Loving Animals* (2011), Kathy Rudy develops an alternative to conventional animal ethics—an ethics based on affect. By taking advantage of narratives of love for animals, it is Rudy's understanding that such narratives ultimately transform societal values and thus practices regarding other animals. But Rudy rejects the power of emotions: sentimentality, empathy, and compassion. It is this simple distinction between the sentimental as superficial and love as deep—ultimately the distinction between emotion and affect—that I am critical of, even though I feel the seduction of Rudy's somewhat naive desire for the transformative power of love. But love is rarely pure or innocent, and does not need to be good. Feminist theory has taught us that love relationships are always part of power relations (Ahmed 2004; Oliver 2009). In this respect, then, there is always a certain amount of selfishness and perhaps even wickedness in care and in love (Haraway 2008a; Holmberg 2011a).

Moreover, in the world's eyes, care for cats goes to excess, and some cat ladies are also lacking in terms of providing care and become the subject of intervention by the community. According to the authorities, such as animal welfare practitioners and behavioral scientists, this is seen as an expression of lack of control, of women losing control. This image is of course in line with a hopelessly ingrained understanding of femininity, a femininity of deficiency, weakness, and negativity (Hirdman 2001). I am nevertheless inclined to interpret the cat lady as a transspecies figure who challenges and subverts, a feline femininity that does not make herself available primarily for other humans in accordance to heteronormative ideals. Instead, she often prioritizes the cat, preferably many. She is independent in her eccentricity, she expresses that she does not care, and her excesses are not mainstream consumptive material. Instead, there are flea market finds and an alternative aesthetic that dominates her home, and you can even see this as an expression of a kind of anti-capitalist critique. But other peculiarities sometimes distinguish cat ladies, who may talk to the animals in specific languages or refrain from feminine coded fashion. Just as other challenging identities or figures such as the "lady" ("tant" in Swedish, Jönsson 2009), the overly feminine "girl" (Österholm 2012), or the "femme-ininity" (Dahl 2012), the cat lady can be said to challenge the cultural norms of femininity as well as animal/human relationships. Analyzing these relationships from a queer feminist intersectional perspective— where heteronormative assumptions about women's reproductive care work are questioned and where non-normative relationships with animals can be seen and used as a feminist critique (McHugh 2012)—can be fruitful. At the same time we must not forget the cats' role. The cat lady is composed; she is a material-semiotic figure, both a challenger of certain hegemonic norms, and one that confirms and reinforces others. Nature and culture, and human and animal, meet, clash, and flinch in the complex traffic of the "zoo." Through the disordering of cat lady relations taking place in homes and cats shelters, humanimal crowding challenges prevailing norms. However, while the cat lady challenges the heteronormative feminine ideal and Western norms for how we should live with pets, she also confirms other norms of human/animal relationships, mainly the discourse of human exceptionalism. Nevertheless, this intersection of species and other social orderings has potential for zoo-chaotic subversion by reclaiming sentimentality and expanding the notion of sociality.

Part III

The promises of crowding in zoocities

6 Beyond crowd control

In mapping the traces of a new urban sociology that is neither reducible to people nor place, my aim has been to show how variegated the spacetimes of cities have become, so that cities need to be grasped as relational entities, and to argue that so intricate are the inter-dependencies between humans and non-humans in urban life, that the social agency of the latter can no longer be ignored.

(Amin 2007: 112)

Recapitulations

When I began this project years ago, I became intrigued by the myriad of more-than-human life taking place in cities. In retrospect, it was a kind of a revelation: before, I saw people and cars, birds and houses, but afterwards, I saw connections and disconnections made up of human-non-human relations, interactions, and assemblages. In the company of and with guidance from influential urban researchers and animal studies scholars, I started to look for ways to investigate this more-than-human everyday life. I interviewed authorities—animal police and welfare officers—as well as cat shelter workers and people who had been subjected to interventions from the authorities. I collected welfare reports and scavenged through printed media for relevant material, watched documentaries and looked at endless numbers of reality shows. In good ethnomethodological fashion, I searched for the instances where conduct was questioned, when the ordinary and normal became foregrounded through the contrasts of norm breaking. I found that, by studying controversies and so-called deviant behavior and bodies, the proper and acceptable human/animal relations came out loud and clear. So far so good. But on top of this, the spatial dimension became increasingly important in interpreting these relations: the city as a defining context, and its streets, parks and homes, played crucial roles in defining, deliberating, and debating human/animal interaction. Thus, the following research question was formulated: What are the multi-species experiences and politics of living in a city? Moreover, I asked: Who is allowed where and under what conditions? What does it mean to become one in a multi-species context of many? And, what does it mean to consider spatial formations and urban politics from the perspective of human/animal relations?

In this chapter, I recapitulate the content of the book, first chapter by chapter, and then as a whole, connecting back to the questions above. I will weave together the threads spread across the book, in an attempt—or rather several attempts—to formulate both some points and some new and important questions. I will reconnect and initiate new discussions using the various theoretical inspirations introduced throughout the book. The approach in this and the following chapter is influenced by Donna Haraway's figure of the cat's cradle: string figures as metaphors that can help in narrating multi-species stories (1994). Depending on how you play around with the strings, different configurations of connections will emerge.

The first chapter sets out to frame the issues brought out in the book, in terms of species, senses, and spaces, and the overall approach of paying thorough sociological attention to the dialectics between subjectivity and form, body and city, within a more-than-human theoretical framework. Chapter 2, "Bodies on the beach," investigates, in considerable detail, the question of "allowability" and multi-species politics of place. Admittedly a limited study of a particular place, it nevertheless lays out a critique of traditional urban studies and its focus on human actors in creating meaning and practices of places. Looking at human/dog interaction and the mobilization of emplaced political action, the chapter shows how emotions and species-specific norms, in/of a multi-species place, play crucial roles in the process of inclusion and exclusion, thus producing allowability. Chapter 3, "Stranger cats," develops these findings by investigating human/cat relations in the city, with a focus on ferality. The title of the chapter refers to the ambiguous role of urban cats, whose presences are defined both by their relationship to humans and by the places they inhabit. On the one hand, they are strangers in the sense of being outsiders: their urban presences are needed in order to understand and experience urban life at all; diversity is needed for urbanism. On the other hand, these diversities are constantly reduced through the measures of control found everywhere, also for cats. Their strangeness is sometimes handled as bodies out of place, as anomalies, and consequently neutralized and removed from sight—by means of rehoming or euthanasia. Thus, the potential of their strangeness in meaning-making and management of homeless and feral cats is reduced by means of the interplay between homeability and ferality, in the decisive methodology of "cat-egorization."

The next section starts off with Chapter 4, "Verminizing," which demonstrates, through an investigation of explanations and assessments of so-called animal hoarding, how the fine line between love and abuse in human animal relations, is played out and negotiated. The focus in the chapter is how animal hoarding is made sense of. First, moving through contemporary explanatory models of animal hoarding, the chapter shows how narratives are soaked with the trope of humanness as distinct and elevated from animality. Losing control and rationality, caring too much, or in the wrong way, contributes to notions of deviance. Second, the chapter investigates sense-making in terms of how the interplay of knowing and sensing works in relation to the phenomenon of urban animal hoarding assessment, by deploying the framework of "sensuous governance." The term refers to the ways in which this sensing as a means of knowledge gets involved in power relations and the method of management: knowledge, senses, and power go hand in hand. Thus, the dimension of species adds to the long-lasting sociological interest in sensing as a mode of knowing about our environment. But, above all, the main contribution of this chapter is the development of earlier chapters on the slippery boundaries between pet and vermin, by the zooming in on the process of emplaced "verminizing" of human and non-human actors in the context of urban animal hoarding. The last empirical chapter, Chapter 5, concerns "Feline femininity." It builds on previous analyses of animal hoarding, and develops the framework of intersectionality in an attempt to understand the different power regimes at play in representations of more or less "crazy cat ladies." Far from all of the cat ladies are considered hoarders; it is not the number of feline companions but the nature of the relationships, that are decisive. Consequently, the various emplacements of species and gender, but also of sexuality and class, point to the power of cultural norms in everyday life, of human-ness as well as of animality and felinity. These relationships are analyzed from a feminist intersectional perspective, where heteronormative assumptions about women's reproductive care work is questioned, and where non-normative relationships with cats can be seen and used as a feminist critique. Moreover, simultaneously keeping an eye on the parallel transformation of the cat, in terms of degrees of "pet-ness," inscribes the analysis in a relational approach to species, gender, sexuality, and class—in short, in an intersectional perspective. Moreover, the home and the neighborhood are performed in quite different ways depending on the interaction taking place and the story being told, and it becomes clear how the animaling of woman/cat relations through feline femininity is an emplaced process.

Together, the chapters highlight the diversity in which more-than-human urban relations are formed, regulated and experienced. In all the cases, however, species, spaces, and senses intersect, and the analyses point at both the effects and limitations of contemporary hegemonies, and at alternative modes of urban politics and crowd control, a matter I will develop below.

Reflections

How can one possibly write about humanimal encounters, in ways that capture the complexities of more-than-human worldlings, in the urban terrain? Can I ever say that I have managed? Lefebvre reflects on the impossibility of writing the everyday life, of capturing time, space, and meaning-making through the dialectical process of symbolic systems and material reality, without reducing the everyday to a frozen and objectified projection (1971: 8). Lefebvre means that the objects are never static, since they move, become transformed, shrink, or expand as we try to capture them through sociological inquiry. How can we ever capture—in text—the motion of time, place, and meaning, dialectically producing a multitude of everyday experiences? Depending on the position and movement of the object in focus, different meanings and experiences emerge. Nevertheless, there is a need to try, to address the everyday as, "the dialectical interaction that is the inevitable starting point for the realization of the impossible" (Lefebvre 1971:

18). Translated to the present case, Lefebvre's discussion of the relation between reality and truth presents the dilemma of writing about humanimal experiences in ways that avoid reductionism, but are still objective enough to be considered scholarly.

To this end, I have used the figure, the "virtual object" (Lefebvre 2003 [1970]: 5), of humanimal crowding, as a methodological and analytical device. A virtual object is, in Lefebvre's terms, a way of capturing the "recurrances," the repetitions occurring in everyday life. It is neither an object that exists unexplored prior to the investigation, nor is it a construction of the researcher. It is more of a "possible object," and thus one that can be hypothesized to account for the complex processes and actions of everyday life (Lefebvre 2003 [1970]: 5). In this trajectory, crowding in zoocities aims at capturing and analyzing some of the complexities described above, bringing together the more-than-human approach, with classical and contemporary urban sociology. Set in motion, the figure is used both empirically—what is going on in living cities—and analytically: contrasting the abstracted figure of crowding in zoocities, with controversial cases regarding unleashed dogs, feral cats, urban animal hoarding, and crazy cat ladies, certain dimensions will be brought out, while others, undoubtedly, remain unexplored.

The fact that some aspects remain unexplored is due to my own bias: I am no doubt guilty of a Western perspective. Admittedly, the manifestations of the order/disorder, purity/dirt dialectics that come to the fore are not universally valid. Alice Hovorka aptly notes, in an article on urban animals in Africa, how the role and agency of urban livestock have escaped the gaze of social scientists (2008). Empirically, her study of human/chicken relations in Greater Gaborone, Botswana, show how closely these relations have been involved in processes of urbanization as well as social structures and dynamics, where "they transgress an urban imaginary that deems them out-of-place, thus challenging human notions of modernity and constructions of urban space" (Hovorka 2008: 110). It is not clear whether this urban, modern imaginary is a Western construct, but it surely operates also in the global South, however in somewhat different ways. As an interesting note, chickens are finding their ways into back yards and balconies also in Western cities, as the urban farming movement has invented purpose-made urban chicken coops (Back yard chickens 2014).

Another possibly unexplored aspect is whether this book really is looking at multi-species urban politics or is a study of humans, cats, and dogs. Or, even worse, of human perceptions of cats and dogs. I wish to argue that, from a dialectical, sociological and interactional point of view, zooming in on crowds and crowding in the urban context, it is not a weakness to focus on multiple individuals of a few different species. On the contrary, such an approach makes a much needed contribution to urban and cultural sociology, expanding and "animaling" concepts such as sociality, gender, and place-making. One particular advantage is being able to analyze in detail how such cross-species interactions take place, and to use these analyses to deploy the framework of humanimal crowding, thus moving through the gray zone of the individual/collective dialectics. As Ash Amin states in this chapter's introductory quote, urban studies can (should?) no longer limit its scope, non-humans are without doubt constituent of the urban condition (2007). Moreover, the book also, from the deer episode in Chapter 1, through the wildlife interaction in Chapter 2, to the multi-species animal hoarding in Chapter 4, highlights the multi-species embeddedness of the human/cat/dog encounters.

The categories of wild, pet, domestic, vermin, and feral, together with human and animal, are set in motion through the approach used. Linking over to thinking more about the promises of crowding in zoocities, I quote Harlan Weaver on the potentials of embracing uneasiness as a methodological strategy, of becoming uncomfortable "in a good way," as a means of taking up, "the troubles of a more than human intersectionality and find[ing] ways to encourage different kinds of 'becoming in kind'" (Weaver 2013: 706).

Reformulations

The promises of humanimal crowding

For early behavioral and social scientists studying crowding in the 1950s and 1960s, the focus was set mainly on problematic features of being confined to a limited space: stress, poor health, immorality, and more (Ekstam forthcoming). Many of these studies were performed with rodents, thus crowding as a concept has an interspecies history (see Chapter 1). Similarly, sociologists viewed crowds as destructive of individual reason as well as of societal equilibrium. Tarde, for example, noted that crowds, as opposed to "publics," accomplished nothing other than destructive production (Clark 1969). He held on to the idea of the transformative power of crowds, often describing the transformation in psychic terms (lunacy, madness) or animalistic terms (e.g. the crowd as animal or beast). However, as noted by recent scholars (Walker 2013: 229), he also acknowledged that the "group mind" that crowds develop could have positive, social cohesive effects, for example at festivals and mourning processions (Tarde 2013: 233). Tarde thus distinguished between good and bad crowds, admitting that, as a whole, "Crowds, therefore—gatherings, get-togethers, exchanges—are much more beneficial than harmful in displaying sociability" (Tarde 2013: 233). Social science scholars such as Brighenti (2010a), Borch and Knudsen (2013), Drury and Stott (2011), and Waddington (2011) acknowledge these positive and productive aspects of crowd theory and have helped reclaim the concept with a view to understanding contemporary social phenomena like sport events, music festivals, and global protests. But if early theorists such as Tarde came close to overstating binary differences between what he termed the crowd and the public, contemporary theorists may overstate the similarities between crowds and social movements. In my view, crowds may have transformative as well as political effects, however this is not what drives their formation. On the contrary, crowds appear, they emerge, out of a different and affective rationale, for example out of playfulness, anger, or the need for shelter and protection. Moreover, while crowds may entail a shared identity, they do not have a formal structure, no organization, and no outspoken agenda, such as social movements do. They do, however, have some other common features.

As stated in Chapter 1, crowd is a noun signifying an undifferentiated mode or expression, while the verb crowding describes the production of such crowds, through the process of emplacement in a certain social setting. Set in motion through the book, it is now time to evaluate this virtual object, reformulating some common features of crowd/ing. First, crowds are spatial formations, taking place at certain times and in specific places. In Simmel's terms, they enable certain forms of sociation, of socio-emotive-spatial formations of interaction. Second, crowding is about transformative powers. Due to the proximity of bodies in limited spaces, transformations of bodies, senses, and identities emerge, what Simmel would possibly call re-sociations. In humanimal crowds, identities flow between bodies, transforming human-ness into animality, or de-humanizations. Similarly, animals may be humanized, while being given human positions: being in control, taking over, deciding—as well as not acting very animal-like. These transformations are going on in a particular place, thus again pointing at the dialectics between interaction and space. Third, crowds and crowding, as phenomena, diffuse the individual/collective binary (see Chapter 1). Crowds are more than the individual bodies involved; they are spatial *formations* that transform bodies, emotions, cognition, and experiences, through the process of crowding. An important point is that neither crowding nor crowds are essentially about numbers. In the verminization of crowds, numbers matters, but only in space-relative and qualitative terms. Fourth, crowds have political potential and may on the one hand be harshly acted upon. They are carefully policed and may be neutralized by various technologies. On the other hand, the crowd is more and stronger than any one, and the counter-political action of crowding may change normative frameworks. Taken together these dimensions of the crowd, and the more-than-human crowding, exercise certain attractions: becoming in proximity with other bodies of various kinds, and the giving up, to a certain degree, of prescribed individuality through re-sociation of one's species circumscribed subjectivity, offers an alternative mode of everyday life. As such, it is a formation that also poses threats to the individualized consumer society. Some would argue that, in contemporary capitalist culture, people drift in crowd-like formations without will and direction, both virtually and in real life. But this is not the kind of movement and meaning I aim at highlighting. On the contrary, "my" humanimal crowds may well be anti-capitalist in effect.

Crowding, as developed here, is about the spatial formation of a collective in a certain social setting, transforming experiences and identities through formative action, and transgressing categories and individuals. Transformations occur through crowding—not just between individual and collective, but also between human and animal. It is what I called for in the introduction, a dialectical development of Haraway's post-humanist claims through sociological tweaking. Crowding is about the becoming of many through "the dance of relating." Dancing is, as we know, an activity that involves embodied and emplaced movements, proximal rubbing against partners and others, intimacy and temporal transformation of the performance, identity, and experience (Törnqvist 2013). But the humanimal crowding in the context of the book takes place in the city. Thus, "crowding in zoocities" still needs some final considerations.

Zoochaotic subversions

Animal geographer Jennifer Wolch has coined the term "zoõpolis" (1998), through which she formulates an ethically robust future of multi-species living in cities. In Wolch's words, this means living in and acting out a reintegration of humans and other animals, including a more ethical relationship that takes the agency of, and co-habituation with, animals seriously.

> The reintegration of people with animals and nature in zoõpolis can provide urban dwellers with the local, situated, everyday knowledge of animal life required to grasp animal standpoints or ways of being in the world, to inter-act with them accordingly in particular contexts, and to motivate political action necessary to protect their autonomy as subjects and their life spaces. [. . .] While based in everyday practice like the bioregional paradigm, the renaturalization or zoõpolis model differs in including animals and nature in the metropolis rather than relying on an anti-urban spatial fix like small scale communalism.
>
> (Wolch 1998: 124)

Wolch polarizes urban life—as full of promises for everyone, human and non-human alike—with more back-to-nature approaches that deny such potentials. Deep ecology, for example, is often used as an example of an approach that views the city as utterly unsustainable. In Wolch's perspective, the city is already packed with "nature," and the proximity of, and potential interaction with, other animals, may lead to more robust, ethical human/animal relations (Wolch 1998, 2002). Haraway has built on this notion, but in her version the zoõpolis provides a more disordered site, or rather a configuration, of multi-species entanglements, and pro-vides promises of better, more response-able futures in the Anthropocene (2011). In this trajectory, Deborah Bird Rose and Thom van Dooren, write compellingly about the ways in which urban places are storied in terms of the memories, mean-ings, movements, and trajectories of other animals. Acknowledging these stories casts serious challenges to the anthropocentric view of everyday life, as well as of urban planning and politics. It means nothing less than a pressing need for an ethical standpoint, much like that of the zoõpolis, of an emerging multi-species sharing, a future of what they term conviviality: "The more-than-human city as a zone of entangled lives and deaths is an understanding yet to be fully realized" (van Dooren and Rose 2012: 19).

Steven Hinchliffe and Sarah Whatmore (2006), however, take this term a step further, when thinking about the "politics of conviviality," in order to explicate the more-than-human urban relations, less in ethical terms, but as political strug-gles. This means that they look for good examples, as they are already played out, or enacted, as "living cities" (2006). The authors argue for a view of urban politics as a "more-than-human affair" (2006: 124), which is not only about co-habitations or sharing:

> We now want to argue that just as city living is not unaffected by this multiplicity and indefinable variety of inhabitations, so inhabitants are not unaffected by city living. Indeed, we want to suggest that nonhumans don't just exist in cities, precariously clinging to the towers and edifices of modernity, but potentially shape and are shaped by their urban relations.
>
> (Hinchliffe and Whatmore 2006: 127)

Like Rose and van Dooren, the authors stress the agency of animals in building cities and producing urban politics. Hinchliffe and Whatmore ultimately wish to make cities livable to all inhabitants, while acknowledging that neither human nor non-human subjects are fixed prior to political interventions, but are formed through their urban becomings. What, then, is a politics of conviviality? In one sense, the authors come close to Wolch's zoöpolis and mean essentially a politics of living (well) together. But, on the other hand, the approach offers a coupling of agonistic politics with an epistemological imperative, in that the politics of knowledge, and who is the knowable subject, is challenged. The categories of lay and expert, research subject and object, become contested and blurred. In the end, corporeal generosity as an ethical guideline, is advocated:

> In this vein, the politics of conviviality implicated in our analytical shift towards a "living city" demands that attention be paid to the diversity of ecological attachments and heterogeneous associations through which the politics of urban nature is fabricated, rather than reading the political ecology of the city off a priori or abstract social divisions.
>
> (Hinchliffe and Whatmore 2006: 135)

The embracing of the heterogeneity of inhabitants, of "accommodation of difference" (2006: 125), is thus advocated, while acknowledging that such diversity will no doubt lead to political struggles.

When moving on to reformulating the concept of "zoocities," I bring with me the notion of more-than-human politics as an agonistic affair of living cities. In Chantal Mouffe's view, this notion derives from a firm critique of liberal democracy, in which plural positions get reduced—via mechanisms of rationalization and consensus—to a single viewpoint (Mouffe 1999). As an alternative, she claims that the dichotomies of we/them and private/public need to be broken down, and animated debates across these and other dichotomies facilitated, in order to allow (that wonderful word!) for a pluralistic and passionate agonistic approach. Needless to say, other animals are not attributed rights as political actors. In addition, Mouffe's altogether discursive approach is not fully applicable to human/animal relations. However, as Sue Donaldson and Will Kymlicka convincingly argue in their relational, political animal rights approach, other animals can be viewed as either citizens or denizens, based on their relationship to humans (2011: 13), and thus as political agents with certain rights. Reconnecting with the more-than-human urban politics explored throughout this book, I thus see some clear benefits with the agonistic approach, in which the political potentials are opened up also between species. Crowding processes are ideal platforms for such agonistic encounters to emerge, be played out, and highlighted, and where humans and animals alike may explore—in relation to differences in response-ability— ways of dealing with heterogeneity in non-conforming ways.

With the term "zoocities," I stress the dialectics between the disordered, anarchistic, and messy, and the ordered, hygienic and clean, as two sides of the same coin; take away one side and the other diminishes too (see Chapter 2). This approach, then, is less a utopian/dystopian vision than an analytical device—a conceptualization of what is actually already going on, that brings out hegemonic narratives through attending to marginal stories. Thus, the concept comes close to Whatmore and Hinchliffe's "living city." Behind, beneath and above all the order- ing of humanimal relations taking place, paying close attention to practice reveals a rather diverse, disordered, and perhaps disturbing experience. In the words of Donna Haraway, these "leaks and eddies might help open passages for a praxis of care and response—response-ability—in ongoing multispecies worlding on a wounded terra" (2012: 302). Thus, the promises of crowding in zoocities lie not so much in peaceful reconciliation, but in the potential ruptures and leakages that the agonistic heterogeneity poses to the ideals of hegemonic purity.

7 Open endings

Co-authored with Katja Aglert

This chapter departs from and further engages with the central topics of the book, opening up discussions of what the findings could potentially mean in terms of multi-species futures, urban ecologies, humanimal ethics, and embracing hetero- geneity. In a conversation with artist Katja Aglert, she and I zoom in on some of the problems raised in the book, our aim being to linger a little longer with the trouble. Aglert has been working with a similar complex of problems—constructions of vermin, boundaries of wild/domestic, nature/culture, native/invasive—yet in other contexts, and more importantly, from the perspective of contemporary art (2012). The idea behind collaborating with Aglert for this chapter—her coming from the inherently interdisciplinary field of contemporary art, and my working with prob- lematizing and illuminating the in-between spaces of dialectical processes—is for the artist's perspective to vitalize the academic discourse and show its flaws as well as strengths. Thus, our collaboration is clearly in line with the interdisciplin- ary trajectory of the book. Animals are, as pointed out, "undisciplined," and the exploration of human/animal relations is a matter that transgresses disciplinary boundaries (Segerdahl 2011). Moreover, we wish to break with the genre of the traditional "conclusion" chapter, which often closes the arguments. Instead, we will provide a chapter that opens doors to new directions, points at potential new areas of use, and widens the scope of the initial questions. In line with this idea, the chapter is structured as a conversation. In reality, the conversations on which the chapter is based took place on several different occasions, both in the office, at cafes, and at home. The recorded conversations in total cover two and a half hours (much more unrecorded conversation went on, for sure) of transcribed data, edited by us both. The result is a joint exploration of the themes of the book, bringing them into "open endings."

Leakage is what's normal

Katja: The subtitle of this book is zoocities, should we start by talking about that term?

Tora: I think it aptly illustrates the emerging problems of the book: in everyday language, zoo refers to chaos and disorder, and to something that is sim- ply disorganized. At the same time, it is a prefix to describe something that has to do with animals.

K: But also, and this is what you and I have discussed earlier, that it refers to a certain place, an organization of a place—that is, the zoo. And there we have all species neatly selected and ordered instead, right.

T: I think that this illustrates something important about the order/disorder complex that reappears throughout the book. That the more you try to arrange and standardize, be it through rules and regulations or through more implicit norms, the more it is revealed how disordered things really are.

K: Could it be that the categorization itself—the ordering, the sorting of, for example, species, if we're to use the example as in the zoo-situation—highlights precisely our desire as humans to define where species "start and end," so to speak, and that it also makes it all too obvious how impossible that very task is?

T: Exactly, it becomes a kind of taxonomy that is firmly constructed. But that's also the available route we have for coping with the complex system that we call nature.

K: In *On Invasive Ground*, one of my previous art projects, the research takes off from the histories of diverse species of flora on the island of Suomenlinna outside Helsinki, Finland. In short, the project investigates ideas behind classifying plants as native or invasive, and furthermore how this connects to shaping national identity. One of my main interests is in all the gaps and contradictions this whole attempt at ordering and selecting involves. The flora archivist I interview speaks of some plants as being taxonomically complicated. She uses docks (the plant genus *Rumex*) as an example, since they reproduce across species boundaries and suddenly you can't taxonomically define them. Now, in your case we're talking about human/animal relations, but my interviewee in *On Invasive Ground* points out a similar problem, which is precisely the underlying chaos that we're discussing.

T: The whole idea behind this botanical taxonomy, well, the animal order too, of course, is precisely that they should not be able to reproduce across species lines. This is the biological definition of a species, that it doesn't reproduce with other species. Or, if it does, the offspring will be sterile, like a zebra-horse or a mule. But even the most scientifically derived taxonomy has its gray zones. Or its liminal spaces, as one might say in relation to this topic.

K: One of the interesting things about Suomenlinna is that it is a world heritage site to be preserved for future generations. With this concept comes strong interests in making it available for tourists, and thereby it has to be historically and taxonomically comprehensible. And when these plants start behaving like chaos pilots, then it becomes a problem precisely in relation to the context that the world heritage constitutes. The fact that you cannot categorize these plants suddenly becomes relevant, right here, in this place. But had it been on the island next door, no one had bothered, they would just have been plants in a landscape.

T: There is something about the practice of starting to sort things out, suddenly it becomes important to make these different categorizations. Your reasoning comes close to Harold Garfinkel's idea of "indexicality," that categorizations are always made meaningful through certain contexts and not others. With this book, we can add spatiality and species to other indexical factors. But one must ask, what kind of order is it that is reproduced? With companion animals or pets, it appears to be more visible, because they are open to categorizations and can be ordered in multiple ways. Then you come along and say that it's the same with some of the plants. And as soon as you start talking about relations and relationalities, what I find so intriguing with Donna Haraway's and others' conceptualization of species per se as a relation. Species themselves constitute and inhabit relations, so to speak, it's always this becoming with. It is so obvious in my examples concerning the dogs and cats, since they're categorized in so many different ways, depending on the socio-spatial context. This points at the existence of liminal creatures, which also enables an opening toward ideas that expand beyond fixed categorizations.

K: Would you say that zoocities includes a critical potential from your side, revealing the attempts to organize the world?

T: There is a critique of the modern obsession, at least in many Western societies, with organizing and following a strict order when relating to other species. But it shows how the codes of conduct are constantly being challenged.

K: This is highlighted through all the examples in this book, and perhaps extra obviously when speaking about numbers, as in the case of cat ladies and hoarders.

T: Exactly! How many animals are you allowed to house in a dwelling, or how many dogs are "many" on a beach? This of course differs from country to country, as well as locally, but I think this is something that we'll see more of. Also general exclusions of pets from rental and social housing.

K: On the other hand, I read recently that a hotel at Stockholm Arlanda airport offers housing for people and their pets during New Year's Eve, in order to get away from the fireworks in the city. Located at the airport it becomes a free zone from all the noise. That's an example of inclusion. But are there any policies regarding numbers of animals allowed?

T: The Swedish law states that nine adult animals is the general limit, but at the same time, which the cat ladies chapter shows, it's not always the exact numbers that are of importance when defining the proper boundaries. It's not the defining marker for conceptualizing a cat lady, in itself a relation between cat and woman. It's also the context: culturally, socially, economically, and legally. I mean, if a breeder has fifty cats or dogs, or a farmer hundreds or even thousands of cows, it doesn't break with dominant norms. Since they're framed within the apparatus of capitalist production, the relations make sense, so to speak.

K: Can we also talk about a certain architectural context for how the relation is framed?

T: Absolutely, this is so true. If you were to rebuild your home to make it fit your cats—or rabbits, as Julie-Anne Smith did—and then sleep in a cat bed yourself, then you'd for sure be called a "crazy cat lady."

K: As someone who's crossed the line for what is considered normal.

T: Exactly. You see it in different reality shows, when an expert comes home because of animal behavior problems, or you have problems with the pet. I saw one episode, in which a rabbit dominated the whole family and ran around and bit people. So a rabbit expert came in and taught the family that this is out of order, a mess, you have to . . .

K: . . . clean it up!

T: Right, rationality and alles in ordnung! Distinctions and different designated places for animal/human encounters. You have dog parks and cat homes, and certain train cars, and right outside this office, the swan pond. But at the same time, I doubt that the distinctions work. There are always leakages. I think leakage is what's normal, but distinctions are the most visible. Well, it's really some kind of chimera.

K: You mean that the chimera is the convention that the leakage is considered what is set apart, while in fact it is what should be considered normal, the dominant order?

T: Yes, I think that this is the common pattern, but it's the distinctions that visualize how dispersed things really are. To think like Foucault: the more we try to separate and organize, the more confused it becomes, because the disciplinary power will never completely succeed.

Liminal creatures and sensuous flow

K: Should we discuss in what ways you focus on the concept of "liminal space" in your chapters?

T: Well, it goes all the way back to the choice of methodology. The ethnomethodological approach is about finding out how actors make sense of their environment. What kind of "methods" are used, what kind of frames and systematizations? How do they go about it? Harold Garfinkel came up with the idea, sent his students out to do things that broke with the norms of conduct, for example to visit their families and behave like a guest, a stranger, in order to see what happens. When norms are broken, you have the best chance of figuring out what is otherwise unreflected practice. In this trajectory, I think the liminal cases are exceptionally good for getting at taken-for-granted and institutionalized patterns of interaction. Take for example the case of urban cats and dogs, as in this book, and you may think that this would be fairly straightforward: people walk their dogs and the cats stay at home or something. But then, when you start digging a little, it seems like urban cats and dogs inhabit highly liminal spaces. Since they're common creatures, formed by their human relations and often living in our company, they are not animal in any ordinary sense, as something other than human, as the other to the human. Rather, they are often seen as family members. But they're not human either. I can't for example take time off from work because my dog is ill. So there are some clear boundaries, but among them there is a kind of liminal space.

K: Right, one of Haraway's motifs for theoretical reasoning has been the cyborg, could pets be considered manifestations of a cyborg?

T: Yes, because they embody the hope of cyborgs. And, in addition, all of Haraway's figurative tropes, all her material-semiotic figures, they have, so to speak . . . they all have a very problematic background and so do the pets. I mean, they are the manifestations of human power and the will to domesticate other species and kind of force them to live close to us. We control their reproduction and their movements, even their looks and behavioral features, through breeding. So, we have power over life and death, over pets, so it's a fairly troublesome relation, to say the least. But, at the same time, there may be liberating aspects, namely the subversive potential in boundary transgressions, which also highlights how humans are also animals.

K: How, what do you mean?

T: You only need to turn on the radio to hear the reoccurring rhetoric of how we stand out from nature, from animals, that we're on top of the great chain of being, or whatever. But we're not, we're also animals. So, the pet figure inhabits both the problematic and the subversive, at once, which in a sense gives us a chance for subversion of the ideology of human exceptionalism. Haraway talks about "companion species" as kin—that could be of any kind really—in order to highlight this complex, and scholars such as Erica Fudge and David Redmalm also write precisely about pets being liminal as revealed through the need to constantly negotiate their place and meaning. Well, we always need to negotiate, that's part of being social, but with regard to these liminal zones, it becomes even more obvious. Our taken-for-granted practices become visible.

K: I think we may also think about liminal space as space for a potential existence beyond dichotomies, a space for equal opportunities, to use a rather basic term. Where power structures can be negotiated from a somewhat common level, a third space, in a sense.

T: Right, a third space! Well, if you think for example about a border zone between nation states, it's a space that's not regulated the same way as other spaces usually are. It becomes undefined, and so much more dangerous in a cultural sense, as Mary Douglas has taught us, and thus new rules emerge and apply. Transgressions in loosely defined spaces become doubly risky, since it threatens the whole system. Thus new rules of harsh kinds may appear, but with little success. Take graffiti as an example. In Stockholm, a zero-tolerance policy emerged in the late 1990s, which only led to higher prevalence. Similarly, the prohibitions against the previously allowed unleashed dogs on the beach (Chapter 2), in practice have little, or even reverse effects in the intended direction.

K: I also think that, since it's a border zone, and potentially open, it makes it possible to deliberate around what kind of order we want for this particular space? If you think about it as a spatial concept, and for example dogs and cats embodying this space, then they can get a powerful and hopeful identity from a power sensitive perspective. It would pose a threat to us, to our hegemony, but it is still an exciting thought. Then it's perhaps interesting to think of these cat ladies, where a different order may have emerged, but I don't know if one could say that it's become more equal or fair. It's so complex, how healthy is it for the cat, and for the human to co-exist on more liminal terms, is it "good" per se? That, you never know, but I find it intriguing since, if you take a spatial perspective, a consequence emerges that, based on societal norms, is viewed as wrong. Abnormal. But if you try to think outside the box, and think about the third space where a different order may appear, that is a real challenge.

T: In parallel, then, Giorgio Agamben talks of the result of the historical separation and hierarchization between humans and animals, as "the open." You may think that it's equivalent to a kind of liminal zone of possibilities. I think of it as a potential for action, but also as a chasm we can't cross without unforeseeable risks, and that we constantly, through what Agamben calls the "anthropological machine" (or rather in the plural—in science, literature, philosophy, etc.), produce the category of "the human" in relation to the ultimate other, "the animal." And what would happen if we stopped? Would we fall in to the chasm or what? The open is still largely unexplored. In Agamben's texts, it becomes clear that he is not really interested in animals at all, but he's interested in what happens when human subjects become transformed into or are reduced really to animals; when individuals or groups no longer can claim the position. And his critique, actually, is that as long as "human" is something that you need to qualify for—rehearsed through the anthropological machine—then there will always be categories of humans that don't really make it and who thus may be treated like animals. And this is of course the core of the problem, something that is not however Agamben's dilemma: Why is it so bad to be treated like an animal? Because they are treated badly! The workings, or should we say disworkings, of the anthropological machine become so obvious in the examples in this book, when you get close enough to this boundary, or too close to the open, you risk becoming less than human. And sometimes explicitly animalized, as "animal like." Because becoming de-humanized and animalized are, strictly speaking, not the same, although related and continuous processes. Another example is the book that you recommended, about the people living in the tunnels beneath New York City.

K: *The Mole People* [Toth 1993].

T: It's exactly that danger, it's somehow the ultimate threat, to be . . .

K: . . . reduced . . .

T: . . . to animals.

K: Degraded. But, considering the core question of this book, the multi-species experiences and politics of living in a city, are you interested in making us reconsider? Do you, in the best of worlds, wish that the gradations didn't exist? That it was simply, I know it's a utopian idea, that we wouldn't think of differences but that it was all about individuals no matter how or what you are, so to speak.

T: We'll never get away from categorizations. There is so much input all the time that, somehow, we need to sort out what we experience in everyday life. So, we'll never get away from categorizations, and I'm not even sure it's desirable either; but to get these orderings rocking, at least a little bit, and to question ourselves on the issue of values and hierarchies. I still believe that moving the human a little bit from the center and focusing more on the relational and dialectical, we will actually shift hierarchies a little.

K: Emotions are interesting in this context, I think that emotions are so—if we talk about the rational as one of the qualities that define and prove our humanness—then emotions are something that make us more animal-like in the sense that we lose our rationality. The artist Oleg Kulik has often taken the role of a dog in his performances, in order to explicitly position himself outside the human horizon. As far as I understand, he does this to reflect a kind of crisis in contemporary culture where, according to Kulik, far too sophisticated speech creates barriers between people. In his interpretation, an appropriation of the dog's behavior pinpoints a kind of communicative potential. I don't see this example as a degradation of humanness, but rather that the liminal creature that the dog embodies may represent a possible escape route.

T: I love that, the dog as hope for better futures; Haraway would agree indeed! I'd like to reconnect to what you said about emotions versus rationality, because it's not as simple as that either, since some emotions are more elevated, and perceived as more human. Empathy for example. Or grief, or existential angst, there are a number of these higher emotions, in contrast to aggression or fear that seem to belong to the "lower" and more bodily affects. Roughly speaking, this is the pattern, but of course it has a lot to do with the context, too.

K: And you talk for example about the notion of senses, in Chapter 5.

T: Exactly, and I think that's what's so great about the cases that I've investigated, that then and there it gets revealed, emotions are permitted to take place. I mean, we experience a multitude of emotions in all social settings and contexts, but we can't always let them out. We have to mask our emotions, most typically perhaps in the academic environment. But when interacting with, in relations to, other animals, the corporeal dimension becomes more important, more profound, and so I think that emotions and affects may flow and be somewhat better accommodated. That there is something going on between bodies, that allows these flows, just like there is in any social relation really, that is accommodating enough.

K: If we'd've had a dog here in this room, then probably one of us would have him/her in the lap now. But what if people started climbing up into one another's laps when they sat down to talk . . . Most probably this is what we used to do ages ago, but then we became more and more separate with civilization. The animals thus serve a function by becoming a tool for us, for channeling our emotions. And thus we face the issue of power, animals rarely exist just because, but pets exist because we have chosen to allow them into our sphere, and to serve us and our purposes. There are stories of stray dogs who approach a person, who then takes on the dog, and then they live happily ever after. Although these stories point out that initially the animals rarely chose their human companion. But, in general, are there any ways of not keeping dogs and cats more or less as slaves, not expecting them for example to affirm you at every given moment, to accept being hugged and to welcome you home every day?

T: Well, all these images, these stereotypes. But there is something about stereotypes, though, that there is often something there, some truth behind them.

K: But if you try to take the dog's or cat's perspective, then, what's the price they have to pay?

T: This is what constantly escapes our attention. It's really difficult to change perspective. But the way I think about emotions, that they're not only something awoken in us, to be channeled out, but flows that are produced and emerge within encounters. So, it's not only coming from inside. See what I mean?

K: But the dog and the cat, they have feelings too, right?

T: That's what I mean. Like for example the dog chapter shows, it's so obviously the dogs' movements and emotions together with people that . . . all kinds of feelings can be created in these encounters: someone gets frightened, another gets happy, a third becomes playful, a fourth experiences panic, and a fifth is just annoyed or angry. So, there is such rich variation and opportunity somehow.

K: The dog as hope for better futures, indeed!

Staying with a messier order

T: What I've tried to demonstrate through the straightforward examples in the book is that, even when looking at human/cat relations only, it's such an incredibly multifaceted and complex situation. Interaction with other species continuously changes our perspective. I say "our," meaning human perspectives in a way, in the plural. How we engage with the city, but also how places are re-shaped depending on these multi-species encounters and urban experiences. If I think about your field, art, I think of Bryndís Snæbjörnsdóttir and Mark Wilson, and their work on animals in the city, and how these encounters shape places and emotions. Their project *Radio Animal* [2009] reveals stories about this through the voices of interviewed locals telling anecdotes about human/animal encounters. But I also think about how art can be deployed to re-shape and challenge experiences and feelings of such multi-species encounters in various places.

K: Thinking about Snæbjörnsdóttir and Wilson, they've looked specifically at the urban garden as space for multi-species encounters. When you talk about urban experiences, are you making any distinctions between public, semi-public, or private?

T: No, I haven't done that.

K: What are your thoughts about that, is there such a focus?

T: No there isn't.

K: They intertwine.

T: Yes, they are intertwined, as you might have noticed.

K: What do you think about the fact that they're entangled, is the distinction needed or is the convergence a narrative in itself? A perspective that is important to claim, perhaps?

T: I think so. There are different schools, of course, but if you focus on the hoarders for example, thinking about private and public, or home versus city, is not fruitful. Then the crucial fact that the dwelling is also in the city and a specific neighborhood disappears. Obviously, the experiences and manifestations will be very different depending on whether the setting is rural or urban. So even if the relations are enacted in the dwelling, this dwelling is located in an urban context, which frames the encounters and the emplacements. Take, for example, cats and the management of cats, it's also a question of how you produce the idea of the housing and the home in an urban context. So it's connected and can't be separated other than analytically. Sophie Watson [2006] writes that the public and the private may rather be seen as mutually constitutive.

K: I think that this also becomes an issue as soon as an individual acts in ways that in any way affect the boundaries between what we consider as home and as public. If a leakage emerges there, then suddenly it becomes a public affair, something that I find is connected to the order you're talking about. In a way you can ask yourself: Why should anyone mind how many cats I have at home? But it's because of the boundary transgression that it suddenly becomes an issue to be handled. Just like the dogs on the beach, there are similar connotations but it looks somewhat different.

T: What do you mean?

K: I think that there's this collective idea about what to do on a beach, it's not private for each and everyone. It is a definite territory called "the beach," where certain actions are anticipated and accepted and not others, whether you are a dog, bird, or human. There are rules for us irrespective of species and if you break these rules it suddenly becomes a public concern.

T: Right, so regarding boundaries, whether they're physical or more socially formulated, they're so immensely crucial to understanding these emplaced and species-specific experiences. In that regard, I also want to talk about the public art project that you brought up earlier, Paul McCarthy's dog poop.

K: *Complex Pile* [2013], a more than 15-meter-high inflatable pile of shit, installed in a park in Hong Kong.

T: What I find so exciting is that he uses the public space in order to highlight an issue that's so amazingly small and commonplace. One of these bothers that are brought up in every residential association, on every block, and in all urban public places. Everywhere, the small piles of feces from cats and dogs become huge matters. In reality, one might think it's a tiny issue, compared to poverty, violence, and environmental pollution, and all kinds of other stuff that could be dealt with instead. But with this giant pile of shit, I think McCarthy makes a very concrete contribution to that debate.

K: Yes, I read somewhere that the strategy of enlarging the issue, the artistic approach that McCarthy regularly uses, when you blow up the problem to absurd dimensions it becomes . . .

T: . . . smaller.

K: It is also significant for how small it really is, and in this case, it is even literally inflated. I saw a picture of it when it was broken, it was lying there without air, and yes, the issue was kind of pooped. *Complex Pile*, the title is definitely spot on.

T: In cities it's a matter of keeping public space free from dirt, and there are a number of technologies and measurements in place, to clean up these piles of poop. It's complex, right, and it also has a very long history of hygienization of the public space, that it should be clean, also preferably from contagion. Matthew Gandy speaks of the "bacteriological city," really one can say the anti-bacteriological. With civilization, we became obsessed with purity.

K: In Berlin for example, you can see where gentrification has taken over, because the streets are free from dog excrement. As soon as you pass the border of what is not yet gentrified, there are piles on the streets. In the neighborhoods of the "new" people, there are these automatized vacuum cleaners that rumble past in the mornings and pick up dog poop, something that is not done in the other neighborhoods.

T: All these technologies or whatever one should call them, the orderly dog owners with their pink plastic dog poop bags in practical designated packings or the machines that clean the street: it's a form of technology that aims at creating a certain order. You have to remember that it's always a matter of negotiations over contingent boundaries, and the outcomes of these negotiations decide that it is dog poop that should be picked up. But we don't jump off the horse to collect its piles when we're out riding, even in densely populated areas. So it's always a matter of, what is and is not viewed as dirt.

K: What do you think, then, is it mere cultural constructions or is there anything factual underlying it? I'm thinking for example of what dogs eat and what horses eat . . .

T: Yes, but still, what is culturally constructed is also material, right. So, there are also aspects such as that and historical facts that we can't ignore. Thus, even if we view it as dirt, culturally, we cannot re-construct that fact in any simple way. As we said, it's a rather tiny issue even if we've enlarged it at this moment, giving it some more space! At the same time, dirt, as Mary Douglas would have it, is so immensely important and interesting when we investigate how orderings work, to understand the worldviews we are dealing with. In David Waltner-Toews' book, *The Origin of Feces*, he demonstrates how we can understand our own as well as other cultures by studying shit and methods of managing it.

K: But we can also learn about the distinction we wish to make about ourselves. I'm thinking about the development of food, how we today eat meat that preferably doesn't resemble dead animals. It's a process we have gone through, at least in the West. In *Hungry City*, Carolyn Steel writes about the ritual carving of the animal body, as an early means of celebrating that its life has been harvested to become food. But now we do all kinds of tricks to hide the origin of a living being. We want to distinguish between animals and animals. There seem to be parallel processes going on while we are becoming more and more sophisticated, and how we shape both food and shit accordingly, and relate to it in similar ways.

T: On the whole, purification and civilization go hand in hand. Norbert Elias has written the history and development of civilization, and it has always been a matter of seclusion. To separate humans from other animals, in order to become more humane. If you look more closely into that process historically, the urban dilemma has been to manage waste. Whether human waste, dead bodies, or garbage, the more "civilized" a place, the more this waste has been managed invisibly. We have gone from disposing of our potty contents on the pavement, to more and more invisible technologies of urban metabolism, hiding what we do not want to see—including meat production. This is one of the more characteristic features of the modern, civilized city.

The relational of the beauty and the beast

K: Thinking of transformations from species to vermin, I'd like to start with an example from my own everyday experience. Our apartment is on the 12th floor in Stockholm and everyone in the building has received a letter, where it's clearly stated that we are not allowed to feed the birds because it attracts rats. But we believe the 12th floor is safe, rats won't come there, so we feed the birds anyway. We suspect that we are the crazy neighbors who have crowds of birds flying around and gathering around our particular balcony. It's probably just a matter of time before complaints will start pouring in, they do make the place fiendishly filthy. Now, we've covered the balcony in plastic so that we can continue feeding them. It's a strategy for allowing continuation, and avoiding public disclosure.

T: Exactly, such a beautiful example of multi-species experiences and politics. On the one hand, it's an attempt to nihilate, and on the other a kind of counter-strategy to allow us to continue doing what we want to do, without offending our neighbors too much. These are the kinds of negotiations that are constantly undertaken around boundaries and around rules and orders, a kind of micro-politics. I think this vermin transformation, it's part of the narratives in all the studies in the book. But in addition I think it's crucial to consider vermin not so much as categories, but rather as processes. What is it that creates, what kinds of dimensions come into the processes of verminizing? Numbers, of course, but it's also a matter of species and the humans involved, species dimensions and place, of course. It's also crucial to attend to the spatial perspective in relation to the making of categories, in relation to where the phenomenon emerges. The cat is over-explicit in that sense, we even name them as house cats, indoor cats, outdoor cats, or wild cats. It's their relationship to humans but also their relations to place and space that define the "cat-egories," including their behavior and personalities. Is it a street cat, then it's feral. In relation to what? Well, it's about their relation

Figure 7.1 Balcony on 12th floor, Stockholm (Photo: Katja Aglert 2013)

to the human. No categories are innocent, so to speak, but are always about various epistemes, as Foucault would put it: the system linking together knowledge, subjectivity, values, and power. This also becomes explicit in the cat example, if it's categorized as a street cat it runs a significantly greater risk of being killed than if it's categorized as a house cat, as someone who is owned. So, both the relation and the place are critical aspects, utterly important. Moreover, the issue of movement, of bodies moving in spaces—for example, I'm thinking about the film I analyze in Chapter 4, a film called *Cat Ladies*, in which the viewer gets to follow one cat lady and her numerous cats, where it's not solely the numbers but also the uncontrollable feline movements that create certain emotions between the film and the spectator. There are cats jumping up and down and kind of moving quickly, thereby breaking with the image of a harmonious home environment. Precisely these emotions that emerge are important dimensions of verminization processes.

K:　I associate this argument with the Australian example of the red fox. English colonizers bring these foxes to Australia without considering these bodies and their "uncontrolled movements," that they will reproduce freely. The consequences are seen as catastrophic some one-hundred years later, since these foxes lack a "natural enemy" in Australia and thus have reproduced to the point where they've now taken over and prey on the native animals. On the other hand, one could ask why this is such a terrible thing? Life on earth has always gone through different phases in which some species become extinct and others not. But, in the end, humanity interferes and intervenes greatly, so perhaps this is the core of the problem. But the mere phenomenon that the world is in constant flux and that we're all affected isn't strange at all. Rather, why should we try to conserve whatever is considered to be Australian? It may sound pragmatic, but still, I think one needs to pose that question.

T:　And this is where nationalism or some kind of territorial identity comes in. The fox is still seen as some kind of intrusion, a kind of fraudulent inhabitant, still some one-hundred years later, as a stranger in Australia. That's so interesting.

K:　We have a similar set of problems in Sweden because of the tree line that is growing further and further north. The red fox is moving with it and taking over large parts of the habitat of the arctic fox, something that is managed by shooting off red foxes and providing space through artificial means. One may ask, what kind of extinction method is that, really? Isn't that a truly fascist method in a sense? There is really no part of nature that humans haven't made decisions over, in one way or another, and these adjustments and corrections are continually being made in various ways. Just because it's about the animals and the ecological balance, we believe there are no ethical problems. But as soon as we decide that a certain species in a certain place is vermin, then we claim the right to eradicate. In urban environments we see a similar situation with for example rats, and I can't foresee the consequences if we were to stop this kind of ordering. But nature handles these situations, creates pests, and so on, so why don't we let nature do the work to a larger extent? But on the other hand, we're also part of nature . . .

T: We are, aren't we, and exactly this idea that both you and I criticize, that nature somehow could be left alone, as if humans weren't also part of nature. On the other hand, where should we draw the line, then, and how much should we interfere? It's also a matter of everything we consume, what we eat and medications we digest, and then our human waste pollutes in one way or another. There are so many species that we don't even consider interacting with. But I'm also thinking about another dimension of this process of verminization, that it may also involve the verminizing of people. The case of animal hoarders shows concretely how norm breaking really makes visible the very ideology of human exceptionalism and what gets to count as the ideal, desirable human-ness. Right these examples, I think it's so obvious that it's a matter of being in control, rationality is also important and some kind of maturity, of adulthood, of course. The infantile, playful and empathic interaction with other animals, and putting them before human subjects, is threatening in some contexts.

K: On the other hand, humans are probably the biggest vermin on earth if you consider that vermin is defined as a species that takes over the native fauna or the native habitat. We've moved forward with some steam, haven't we? I think it becomes obvious when trying to analyze the transformation between species and vermin, and what kinds of traits these identifications include, depending on who you're looking at. What is viewed as rationality in one species is considered vermin behavior in the other.

T: I'd like to add here that there is a tendency within the post-humanist paradigm, as well as the eco-critical one, to degrade and demonize humans. I'd rather see this as a potential for more ethical positions and responsible practices, or response-able as Haraway would put it. Then of course it's such a pity that what's considered a sign of human-ness is still so obviously coded in masculine terms. Sentimentality as connected to femininity is, for example, less worthy. This becomes so transparent in the reality shows that I analyze in Chapter 4, how the disciplining of people who hoard animals, and have too many, is about regaining control, letting go of emotional bonds and becoming more rational. I believe that a more intersectional perspective, of course including species and place, is well worth continuing with.

K: I also believe we need to give up our search for harmony. Last night, I watched a program on the magic of plants or something, with David Attenborough, where he spoke about the diversity and there seemed to be a fundamental narrative of . . . I mean, I also want to see all these amazing plants in the world, but it's a matter of how to formulate the motivation, the motives for keeping that diversity. Of course all these flowers should be there, because we wish for some kind of chaos, of variety in the world, of co-evolution and co-habitation, but it's not incontestably a matter of a homogeneously happy world. This is probably where the dilemma is, this underlying narrative of a kind of harmonious happiness. But to affirm the confrontation, the negotiation must be part of co-existence and it doesn't have to be about me getting my opinion across. I have to make compromises on the road to co-existence. Chantal Mouffe and her model of "agonistic pluralism" and the importance of dissensus in a democratic world puts this into words perfectly.

T: I agree in that it easily spills over into a neo-liberal discourse of diversity that, coming from different cultures is so great, but the discourse of diversity is also in itself extremely homogenizing. It also departs from the harmonious, living side by side, learning from one another and it's thus a discourse that to a large extent ignores power relations.

Crowding: re-ordering urban politics

T: In the research literature on housing, crowding has been discussed in terms of problems. It's about crowded housing creating stress, it's about crowded and dense cities, that transportation is crowded, and how these conditions continuously demand our management. At the same time, there are many positive aspects, that's in part why many of us, including other animals, choose to live in cities. I believe one of the foremost promises of the concept of crowding and the analyses in these terms, when discussing urban human/animal relations, is exactly this point, that we're forced to rub up against one another, and this forms our experiences and perspectives in certain ways. There is something promising about being forced to or choosing to encounter others, strangers, all the time. It is this friction, rather than diversity, because if you talk about diversity you're still stuck in schemes of different categories, of taxonomies. But if you instead think through multi-species crowding, that conceptual frame inhabits an idea of the impossibility of definitely distinguishing between categories as well as species and individuals. Crowds are more than the sum of individuals, and thus transgress the collective/individual dualism. It becomes something else. Crowds/crowding can take different routes, there is something hopeful in that. Crowds also have the ability to challenge power and hegemonies, through the creation of counter-power.

K: But if you consider the public protests and that situation. It's a telling example of how the crowding concept really gets tested, as a method through demonstration as a form, in order to challenge the order coming from the minority in charge. But how well do these crowds really succeed? It seems like most of the examples we see today, at least, so far haven't had the impact they potentially could have had. Whether we're talking about a flock of birds or a flock of people, kind of.

T: What's interesting is rather the non-intentional action and meaning, when we think of the masse, the crowd, we tend to think about political manifestations or the like, which are more akin to social movements with certain specific aims for the action taken.

K: This makes me think of a different example, the barbecuing Turkish families in Berlin. The parks have gained a new function for Turkish families, with aunts and cousins, grandparents and children hanging out in the park all day and cooking. It smells of barbecue all over in the summer season. This is an example of crowding, where it has changed, old norms have given way to new ones in the urban space.

T: Right, because a certain proximity is created, of certain bodies in a certain place, so to speak, and thus the meaning and use of that place changes, the overall organization of bodies. But it's also because it's about continuous and repetitive action—I think of the dogs on the beach in Chapter 2, who return to the beach over and over again, thus contributing to and becoming with the rhythm of the urban space, as Lefebvre wrote, rather than unique and temporary events that take place once and not necessarily ever again. The concept of crowding itself has a temporal aspect, and it is this continuity, insistence, that in the end also can become a subversive force.

Impossible knowledge as means of becoming with

T: The point is also that, if you say that there exists something that is a cat, then in a way, you diminish everything that is not cat. I'm not saying that there are no biological cats and dogs, or biological humans for that matter, I hope we're clear on that and can move beyond such discussions, which were hot in the 1990s. Funnily enough, now the same debate has returned with the "new feminist materialism," but from the opposite direction, so to speak, claiming that there has been too much emphasis on discursive and cultural matters. But matter also matters, as Karen Barad would have put it. Still I believe that it's impossible to refer becoming with or becoming wordly, as we're talking about it here with reference to Haraway, to biological bodies. Instead it's both a matter of the continuous becoming of identities, and the evolution-based physical becoming, and the intertwinement of these processes. There are many processes going on at the same time, subtly embedded in one another across species boundaries. But that doesn't mean that all species all over the world are connected to one another, all the time. There is a certain, critical situatedness that needs to be addressed. Becoming with is always taking place.

K: But the simplified idea that we can easily distinguish between humans and animals has had such a huge impact. Why aren't there more people out there who are interested in complicating that relationship? Is it because it's connected to benefits, that privilege, and that if we start to problematize the way we treat animals, we'll get into serious trouble?

T: Absolutely, and this is obvious also with human rights movements and women's rights movements, we have this history, which is a rather late development toward citizenship that includes people other than privileged white men. But the one who belongs to the norm can't really see and reflect on it, right. A colleague of mine wrote about male surgeons and asked what it's like to be a man and doctor. Another colleague ran into problems when asking about experiences of whiteness in contemporary Sweden. It's like a nonsense question to white people, and almost impossible to answer. If you ask someone out there what it's like to be human, you'll probably have the same kind of trouble formulating something comprehensible. If you belong to the privileged norm, it's difficult to critically reflect on these experiences, to get at the embodied and emplaced subjectivity. What's needed, then, is a different angle, a social scientific or artistic or similar perspective, a critical context that enables a slight movement of one's own positions in order to highlight the normative, the normal. It's also important from my position as a sociologist and social scientist, who often have a critical or constructivist perspective on reality and society and social orders, to be alert to the boundaries being drawn, to think about terms like society, culture, and nature, as not something totally given. As if animals are not constructed too, just as human-ness constantly is.

K: I'd like to address the problem of representation here. As you discuss in Chapter 1, you see it as something of a mission bound to fail, in relation to the possibility of taking the animals' perspective as part of your analyses. Well, it's obvious that we can't take the perspective of a dog, for example, and why would we be able to do that, when we clearly can't take anyone else's perspective, really. Just as little as we can embody another animal's perspective, it's equally difficult to embody a fellow human's perspective. I can neither take a stick insect's perspective nor take yours. Where individuals are concerned, there are as many perspectives as there are individuals.

T: But at the same time, this doesn't serve us well if we want to do research; then we somehow have to make a decision. I also rely on feminist epistemologies that claim the importance of situated knowledges, and not just the individual position, that I am white and middle-class or whatever. It's more a matter of how one can create certain knowledge, from collective and relational positions, that can actually be acquired. Otherwise, no research would ever be done on, for example, marginalized groups in society who are perhaps not represented at the universities, like the precariat. It's somehow one's duty as a social scientist to also think from the margins.

K: One of the strategies in the field of art, of dealing with the issue of representation of the other, has been to also make the sender, that is the artist, transparent, and visible as part of the manifestation. Gitte Villesen is an artist that developed this method in interesting ways over the years. You, on the other hand, are acting in an academic field, and say that you have to fixate some issues, but still you need not give definite answers, right? I see some methodological convergences in that we, while trying to formulate the problem, realize that it doesn't let itself be formulated, and that this is okay. Like with Haraway's notion of staying with the trouble, because it's the only available route regarding this dilemma of representation, there are no clear-cut answers.

T: But you make the best you can out of it. This is a constant ongoing discussion in animal studies too, how can we develop our methods? I have a rather traditional approach for this book, and that's what I am left with at this moment. But I still try to put relationality into focus, from the positions available.

K: Perhaps that's the whole point, that there doesn't need to be a final consensus, but that it's the chaos, the leakage, just like you also say, that's normal. Perhaps it is by developing these questions that one can have an entirely different goal for what one is trying to accomplish.

***

While we go on and on about the problems of representation and the dilemmas of speaking for the subaltern, Ronja the dog suddenly looks at us, yawns, and leaves the room . . .

Bibliography

Acampora, R. (2006) *Corporal Compassion. Animal Ethics and Philosophy of Body*, University of Pittsburgh Press: Pittsburgh, PA.

AETV (2011) *Hoarders*, season 4 (episode 9, Stacey), retrieved from www.aetv.com/hoarders/video/stacey-roi (accessed December 18, 2014).

AETV (2012) *Hoarders*, season 3 (episode 19, Kathy and Gary), retrieved from www.aetv.com/hoarders/video/kathy-gary (accessed December 18, 2014).

Agamben, G. (2004) *The Open. Man and Animal*, Stanford University Press: Stanford.

Aglert, K. (2012) *On Invasive Ground* (art project), retrieved from www.katjaaglert.com/work_on_invasive_ground_aglert.html (accessed August 2, 2014).

Aglert, K., Hessler, S. (eds.) (2014) *Winter Event—Antifreeze*, Art and Theory Publishing: Stockholm.

Ahmed, S. (2000) *Strange Encounters. Embodied Others in Post-Coloniality*, Routledge: London.

Ahmed, S. (2004) *The Cultural Politics of Emotion*, Routledge: London.

Amin, A. (2007) "Rethinking urban social," *City*, 11(1): 100–114.

Amin, A., Thrift, N. (2002) *Cities. Reimagining the Urban*, Polity Press: Cambridge.

Andersson, K. (1997) "A walk on the wild side: a critical geography of domestication," *Progress in Human Geography*, 21(4): 463–485.

Andersson, K. (1998) "Animals, science and the spectacle in the city," in Wolch, J., Emel, J. (eds.) *Animal Geographies. Place, Politics, and Identity in the Nature–Culture Borderlines*, Verso: New York: 27–50.

Animal Planet (2012) *Confessions: Animal Hoarding* (episode: Shelley), retrieved from www.animalplanet.com/tv-shows/confessions-animal-hoarding/videos/confessions-animal-hoarding.shelley.html (accessed December 18, 2014).

Animal Planet (2013a) *Confessions: Animal Hoarding* (episode: Jonnie), retrieved from www.animalplanet.com/tv-shows/confessions-animal-hoarding/videos/animal-hoarding-jonnie.htm (accessed December 18, 2014).

Animal Planet (2013b) *Confessions: Animal Hoarding* (episode: Mike), retrieved from www.animalplanet.com/tv-shows/confessions-animal-hoarding/videos/animal-hoarding-mike.htm (accessed December 18, 2014).

Arbetarbladet (2010) (news article).

Aristocats (1970) (film), Walt Disney Productions: Burbank, CA.

Arluke, A. (2006) *Just a Dog. Understanding Animal Cruelty and Ourselves*, Temple University Press: Philadelphia, PA.

Arluke, A., Killeen, C. (2009) *Inside Animal Hoarding: The Barbara Erickson Case*, Purdue University Press: Lafayette, IN.

Arluke, A., Sanders, C.R. (1996) *Regarding Animals*, Temple University Press: Philadelphia, PA.

Arluke, A., Frost, R., Steketee, G., Patronek, G., Luke, C., Messner, E., Nathanson, J., Papazian, M. (2002) "Press reports of animal hoarding," *Society and Animals*, 10(2): 113–135.

Atkins, P. (ed.) (2012) *Animal Cities. Beastly Urban Histories*, Ashgate: Farnham.

Back yard chickens (2014) (website), retrieved from www.backyardchickens.com (accessed December 18, 2014).

Balcom, S., Arluke, A. (2001) "Animal adoption as negotiated order: a comparison of open versus traditional shelter approaches," *Anthrozoös*, 14(3): 135–150.

BBC Surrey (2013) (website), retrieved from http://news.bbc.co.uk/local/surrey/hi/people_and_places/history/newsid_8211000/8211876.stm (accessed December 18, 2014).

Becker, H. (1963) *Outsiders. Studies in the Sociology of Deviance*, The Free Press: New York.

Bennett, J. (2010) *Vibrant Matter. A Political Ecology of Things*, Duke University Press: Durham.

Birke, L. (2002) "Intimate familiarities? Feminism and human–animal studies," *Society & Animals*, 10(4): 429–436.

Birke, L. (2014) "Escaping the maze: wildness and tameness in studying animal behavior," in Marvin, G., McHugh, S. (eds.) *Routledge Handbook of Human–Animal Studies*, Routledge: London: 39–53.

Birke, L., Bryld, M., Lykke, N. (2004) "Animal performances," *Feminist Theory*, 5(2): 167–183.

Björklund, A. (1994) "Farstas kattmamma och vardagshistorien," in Arnstberg, K.-O. (ed.) *Stockholmsbilder. En festskrift*, Info Books: Stockholm: 153–168.

Bloor, D. (1991 [1976]) *Knowledge and Social Imagery*, University of Chicago Press: Chicago, IL.

Blunt, A., Dowling, R. (2006) *Home*, Routledge: London.

Bonas, S., McNicholas, J., Collins, G.M. (2000) "Pets in the network of family relationships: an empirical study," in Podberscek, A.L., Paul, E.S., Serpell, J.A. (eds.) *Companion Animals and Us. Exploring the Relationships Between People and Pets*, Cambridge University Press: Cambridge: 209–236.

Borch, C., Knudsen, B.T. (2013) "Postmodern crowds: re-inventing crowd thinking," *Distinktion. Scandinavian Journal of Social Theory*, 14(2): 109–113.

Borden, I. (2001) *Skateboarding, Space and the City. Architecture and the Body*, Berg: Oxford.

Bourdieu, P. (2001) *Masculine Domination*, Polity Press: Cambridge.

Bowen, J. (2012) *A Street Cat Named Bob*, Hodder & Stoughton: London.

Bowker, G.C., Star, S.L. (2000) *Sorting Things Out. Classification and its Consequences*, MIT Press: Cambridge, MA.

Braidotti, R. (2008) *Transpositions,* Polity Press: Cambridge.

Brandt-Hawley, S. (2003) Petition for writ of mandamus, Superior Court of the State of California, filed 27 May.

Brighenti, A.M. (2010a) "Tarde, Canetti, and Deleuze on crowds and packs," *Journal of Classical Sociology*, 10(4): 291–314.

Brighenti, A.M. (2010b) "At the wall: graffiti writers, urban territoriality, and the public domain," *Space and Culture,* 13(3): 315–332.

Broberg, G. (2004) *Kattens historia* [The history of the cat], Atlantis: Stockholm.

Brown, N. (2003) "Hope against hype—accountability in biopasts, presents and futures," *Science Studies,* 16(2): 3–21.

Brown, S., Fox S., Jaquet, A. (2007) "On the beach: liminal spaces in historical and cultural contexts," *Limina. A Journal of Historical and Cultural Studies*, retrieved from www.limina.arts.uwa.edu.au/previous/special_edition (accessed August 19, 2010).

Bryld, M., Lykke, N. (2001) *Cosmodolphins*, Zed Books: London.

Bull, J. (2014) "Between ticks and people: responding to nearbys and contentments," *Emotion, Space and Society*, 12(1): 73–84.

Bull, J. (ed.) (2011) *Animal Movement. Moving Animals. Essays in Direction, Velocity and Agency in Humanimal Encounters,* Crossroads of Knowledge no. 17, Centre for Gender Research, Uppsala University: Uppsala.

Butler, J. (1990) *Gender Trouble*, Routledge: New York.

Calhoun, J.B. (1962) "Population density and social pathology," *Scientific American*, 206(2): 139–148.

Casey, E. (1993) *Getting Back into Place: Toward a Renewed Understanding of the Place-World*, Indiana University Press: Bloomington, IN.

Cassidy, R. (2007) "Introduction: domestication reconsidered," in Mullin, M., Cassidy, R. (eds.) *Where the Wild Things are Now*, Bloomsbury Academic: New York: 1–25.

Castrodale, L., Bellay, Y.M., Brown, C.M., Cantor, F.L., Gibbins, J.D., Headrick, M.L., Leslie, M.J., MacMahon, K., O'Quin, J.M., Patronek, G.J., Silva, R.A., Wright, J.C., Yu, D.T. (2010) "General public health considerations for responding to animal hoarding cases," *Journal of Environmental Health,* 72(7): 14–18.

Cat Ladies (2008) (film), Chocolate Box Entertainment: Toronto.

Chen, N.N. (1992) "Speaking nearby: a conversation with Trinh T. Minh–ha," *Visual Anthropology Review*, 8(1): 82–91.

Chiu, C. (2009) "Contestation and conformity: street and park skateboarding in New York City public space," *Space and Culture*, 12(1): 25–42.

City Data (2010) Santa Cruz, California, retrieved from www.city-data.com/city/Santa-Cruz-California.html (accessed December 19, 2010).

Clark, T.N. (1969) "Introduction," in Clark, T.N., Tarde, G. (eds.) *Gabriel Tarde on Communication and Social Influence, Selected Papers,* University of Chicago Press: Chicago, IL.

Classen, C., Howes, D., Synnott, A. (1994) *Aroma. The Cultural History of Smell*, Routledge: London.

Coulon, A. (1995) *Ethnomethodology*, Sage: London.

Cuomo, C.J., Gruen, L. (1998) "On puppies and pussies: animals, intimacy and moral distance," in Bar-On, B.A., Ferguson, A. (eds.) *Daring to be Good: Essays in Feminist Ethico-Politics*, Routledge: London: 129–142.

Dahl, U. (2012) "Turning like a femme: figuring critical femininity studies," *NORA*, 20(1): 57–64.

Davies, L. (2011) "Zones of contagion: the Singapore body politic and the body of the street-cat," in Freeman, C., Leane, E., Watt, Y. (eds.) *Considering Animals*, Ashgate: Farnham: 183–198.

Deleuze, G., Guattari, F. (2013) *A Thousand Plateaus. Capitalism and Schizophrenia*, Bloomsbury Press: New York.

Despret, V. (2004) "The body we care for: figures of anthropo-zoo-genesis," *Body & Society*, 10(2–3): 111–134.

Dikeç, M. (2007) *Badlands of the Republic. Space, Politics, and Urban Policy*, Blackwell Publishing: Oxford.

Dion, M., Rockman, A. (1996) *Concrete Jungle*, Juno Books: New York.

Dirke, K. (2007) *De värnlösas vänner. Den svenska djurskyddsrörelsen 1875–1920*, Almqvist & Wiksell International: Stockholm.

Dog Park USA (2010) "Santa Cruz County pet friendly places," retrieved from www. doggoes.com/parks/california/santa-cruz-county (accessed December 9, 2010).

Donaldson, S., Kymlicka, W. (2011) *Zoopolis. A Political Theory of Animal Rights*, Oxford University Press: Oxford.

Donovan, J. (2007) "Animal rights and feminist theory," in Donovan, J., Adams, C.J. (eds.) *The Feminist Care Tradition in Animal Ethics*, Columbia University Press: New York: 58–86.

Douglas, M. (1997 [1966]) *Purity and Danger. An Analysis of Concepts of Pollution and Taboo*, Taylor & Francis Ltd: London.

Driessen, C., Bracke, M.B.M., Copier, M. (2010) "Designing a computer game for pigs; to create a playful interface between animals, science and ethics," paper for Practicing Science and Technology, Performing the Social, conference of the European Association for the Study of Science and Technology, 2–4 September, Trento, Italy.

Drury, J., Stott, C. (2011) "Conceptualizing the crowd in contemporary social science," *Contemporary Social Science*, 6(3): 275–288.

Duyvendak, J.W. (2011) *The Politics of Home. Belonging and Nostalgia in Western Europe and the United States*, Palgrave Macmillan: London.

Ekstam, H. (2013) "Om trångboddhet: Hur storleken på våra bostäder blev ett välfärdsproblem," *Sociologisk Forskning*, 50(3–4): 199–122.

Ekstam, H. (forthcoming) "Residential crowding in a 'distressed' and 'gentrified' neighbourhood. Towards an understanding of crowding in 'gentrified' neighbourhoods," *Housing, Theory & Society*.

Elias, N. (2000) *The Civilizing Process. Sociogenetic and Psychogenetic Investigations*, Blackwell Publishers: Oxford.

Elias, N., Scotson, J.L. (1965) *The Established and the Outsiders. A Sociological Enquiry into Community Problems*, Frank Cass & Co: London.

Emel, J. (1998) "Are you man enough, big and bad enough? Wolf eradication in the US," in Wolch, J., Emel, J. (eds.) *Animal Geographies. Place, Politics and Identity in the Nature–Culture Borderlands*, Verso: New York: 91–116.

Farias, I., Bender, T. (2011) *Urban Assemblages. How Actor–Network Theory Changes Urban Studies*, Routledge: London.

Feld, S. (2005) "Places sensed, senses placed: toward a sensuous epistemology," in Howes, D. (ed.) *Empire of the Senses. The Sensual Culture Reader*, Berg: Oxford and New York: 179–190.

Feld, S., Basso, K. (eds.) (1996) *Senses of Place*, School of American Research Press: Santa Fe, NM.

Fiske, J. (1989) *Reading the Popular*, Taylor & Francis Ltd: London.

FOLF (2005) Friends Of the Lighthouse Field, retrieved from www.folf.org/issues (accessed August 18, 2010).

FOLF (2010) Friends Of the Lighthouse Field, retrieved from www.folf.org/issues (accessed August 18, 2010).

Foucault, M. (1991) "Governmentality," in Burchell, G., Gordon, C., Miller, P. (eds.) *The Foucault Effect: Studies in Governmentality*, University of Chicago Press: Chicago, IL.

Foucault, M. (1998) *The History of Sexuality, Part 1*, Penguin Books: London.

Fox, R. (2006) "Animal behaviours, post-human lives: everyday negotiations of the animal–human divide in pet-keeping," *Social and Cultural Geography*, 7(4): 525–537.

Fox, R., Walsh, K. (2011) "Furry belongings: pets, migration and home," in Bull, J. (ed.) *Animal Movements, Moving Animals. Essays in Direction, Velocity and Agency in*

Humanimal Encounters, Crossroads of Knowledge no. 17, Centre for Gender Research, Uppsala University: 97–117.

Francis, D., Hester, S. (2004) *An Invitation to Ethnomethodology. Language, Society and Interaction*, Sage: London.

Franck, K., Stevens, Q. (2006) *Loose Space. Diversity and Possibility in Urban Life*, Routledge: London.

Franklin, A. (1999) *Animals and Modern Cultures. A Sociology of Human–Animal Relations in Modernity*, Sage: London.

Franklin, A. (2006) "Be (a)ware of the dog. Post-human housing," *Housing, Theory & Society*, 23(3): 137–156.

Franklin, A. (2014) "The adored and the abhorrent: nationalism and feral cats in England and Australia," in Marvin, G., McHugh, S. (eds.) *Routledge Handbook of Human–Animal Studies*, Routledge: London: 139–154.

Franklin, A., White, R. (2001) "Animals and modernity: changing human–animal relations, 1949–98," *Journal of Sociology*, 37(3): 219–238.

Franzén, M. (2002) "A weird politics of place: Sergels torg, Stockholm (round one)," *Urban Studies*, 39(7): 1113–1128.

Freud, S. (1919 [2003]) *The Uncanny*, Penguin Books: London.

Frost, R. (2000) "People who hoard animals," *Psychiatric Times*, 17(4): 25–29.

Frost, R.O., Steketee, G. (2010) *Stuff. Compulsive Hoarding and the Meaning of Things*, Houghton Mifflin Harcourt: Chicago, IL.

Frost, R.O., Patronek, G., Rosenfeld, E. (2011) "A comparison of object and animal hoarding," *Depression & Anxiety*, 28(10): 885–891.

Fudge, E. (2008) *Pets*, Acumen Publishing: Stocksfield.

Gaarder, E. (2011) *Women and the Animal Rights Movement*, Rutgers University Press: New Brunswick, NJ.

Gandy, M. (2003) *Concrete and Clay. Reworking Nature in New York City*, MIT Press: Cambridge, MA.

Gandy, M. (2006a) "Zones of indistinction: bio-political contestations in the urban area," *Cultural Geographies*, 13(4): 487–516.

Gandy, M. (2006b) "Bacteriological city," *Historical Geography*, 34: 14–25.

Garfinkel, H. (1967) *Studies in Ethnomethodology*, Polity Press: Cambridge.

Garfinkel. H. (2002) *Ethnomethodology's Program: Working out Durkheim's Aphorism*, Rowman & Littlefield: Lanham.

Gieryn, T. (2000) "A space for place in sociology," *Annual Review of Sociology*, 26: 463–496.

Giffney, N., Hird, M.J. (eds.) (2008) *Queering the Nonhuman*, Ashgate: Aldershot.

Goffman, E. (1974) *Frame Analysis. An Essay on the Organization of Experience*, Harvard University Press: Cambridge, MA.

Goffman, E. (1990 [1971]) *Stigma. Notes on the Management of a Spoiled Identity*, Penguin: London.

Gove, W.R., Hughes, M., Galle, O.R. (1979) "Overcrowding in the home: an empirical investigation of its possible pathological consequences," *American Sociological Review*, 44(1): 59–80.

Grier, K.C. (2006) *Pets in America. A History*, University of North Carolina Press: Chapel Hill, NC.

Griffiths, H., Poulter, I., Sibley, D. (2000) "Feral cats in the city," in Philo, C., Wilbert, C. (eds.) *Animal Spaces, Beastly Places. New Geographies of Human–Animal Relations*, Routledge: London: 56–70.

Guardian (2011) "Missing Colorado cat found in New York," 15 September.

Halberstam, J. (2011) *The Queer Art of Failure*, Duke University Press: Durham.

Hannerz, E. (2013) *Performing Punk: Subcultural Authentications and the Positioning of the Mainstream*, PhD diss., Uppsala University: Uppsala.

Haraway, D.J. (1991) *Simians, Cyborgs, and Women. The Reinvention of Nature*, Routledge: New York.

Haraway, D.J. (1994) "A game of cat's cradle: science studies, feminist theory, cultural studies," *Configurations*, 2(1): 59–71.

Haraway, D.J. (1997) *Modest_Witness@Second_Millenium. FemaleMan©_Meets_OncoMouse™*, Routledge: New York.

Haraway, D.J. (2003) *Companion Species Manifesto*, Prickly Paradigm Press: Chicago, IL.

Haraway, D.J. (2008a) *When Species Meet*, Minnesota University Press: Minneapolis, MN.

Haraway, D.J. (2008b) "Companion species, mis-recognition, and queer worlding," in Giffney, N., Hird, M.J. (eds.) *Queering the Non/Human*, Ashgate: Aldershot.

Haraway, D.J. (2010) "When species meet: staying with the trouble," *Environment and Planning D. Society & Space*, 28(1): 53–55.

Haraway, D.J. (2011) "Zoöpolis, becoming worldly, and trans-species urban theory: for old cities yet to come," paper presented at Playing Cat's Cradle with Companion Species: The Wellek Lectures, UC Irvine, US.

Haraway, D.J. (2012) "Awash with urine: DES and Premarin© in multi-species response-ability," *WSQ: Women's Studies Quarterly*, 40(1–2): 301–316.

Hayward, E. (2008) "Lessons from a star fish," in Hird, M., Giffney, N. (eds.) *Queering the Nonhuman*, Ashgate: Aldershot: 249–263.

Hayward, E. (2010) "FingeryEyes, impressions of cup corals," *Cultural Anthropology*, 25(4): 577–599.

Hayward, E. (2011) "Sounding out the light: beginnings," in Segerdahl, P. (ed.) *Undisciplined Animals. Invitations to Animal Studies*, Cambridge Scholars Publishing: Cambridge: 159–185.

Hebdige, D. (1979) *Subculture. The Meaning of Style*, Methuen & Co: London.

Heritage, J. (2012) "Epistemics in action: action formation and territories of knowledge," *Research on Language & Social Interaction*, 45: 1–29.

Hester, S., Eglin, P. (1997) *Culture in Action: Studies in Membership Categorization Analysis*, University Press of America: Lanham, MD.

Hinchliffe, S. (2007) *Geographies of Nature. Societies, Environments, Ecologies*, Sage: London.

Hinchliffe, S., Whatmore, S. (2006) "Living cities: towards a politics of conviviality," *Science as Culture*, 15(2): 123–138.

Hinchliffe, S., Bingham, N. (2008) "People, animals and biosecurity in and through cities," in Ali, S.H., Keil, R. (eds.) *Networked Disease*, Blackwell Publishing: Oxford: 214–228.

Hirdman, Y. (2001) *Genus. Om det stabilas föränderliga former*, Liber: Malmö.

Hobson-West, P. (2007) "Beasts and boundaries: an introduction to animals in sociology, science and society," *Qualitative Sociology Review*, 3(1): 23–41.

Holmberg, T. (2005) *Vetenskap på gränsen* [Science on the line], Arkiv Förlag: Lund.

Holmberg, T. (2006) "Hur gör djur? Människor, andra djur och utmaningar för sociologin," *Sociologisk Forskning*, 43(1): 5–20.

Holmberg, T. (2007) *Samtal om biologi*, Skrifter från Centrum för genusvetenskap 4, Uppsala universitet, Uppsala.

Holmberg, T. (2011a) "Controversial connections. Urban studies beyond the human," in Segerdahl, P. (ed.) *Undisciplined Animals. Invitations to Animal Studies*, Cambridge Scholars Publishing: Cambridge: 135–156.

Holmberg, T. (2011b) "Mortal love. Care practices in animal experimentation," *Feminist Theory*, 12(2): 147–163.

Holmberg, T. (2014a) "Wherever I lay my cat? Post-human urban crowding and the meaning of home," in Marvin, G., McHugh, S. (eds.) *Routledge Handbook of Human–Animal Studies*, Routledge: New York: 54–67.

Holmberg, T. (2014b) "Sensuous governance. Assessing urban animal hoarding," *Housing, Theory & Society*, 31(4): 464–479.

Hopper, K. (1990) "Public shelter as 'a hybrid institution': homeless men in historical perspective," *Journal of Social Issues*, 46(4): 13–29.

Hovorka, A.J. (2008) "Transspecies urban theory: chickens in an African city," *Cultural Geographies*, 15(1): 119–141.

Howell, P. (2012) "Between the muzzle and the leash: dog-walking, discipline and the modern city," in Atkins, P. (2012) *Animal Cities. Beastly Urban Histories*, Ashgate: Farnham: 221–241.

Howes, D. (2005) "Introduction," in Howes, D. (ed.) *Empire of the Senses. The Sensual Culture Reader*, Berg: Oxford and New York: 1–17.

Humanimal Group (forthcoming) *Zooethnographies Manifesto*, unpublished manuscript.

Hutson, S., Clapman, D. (1999) *Homelessness: Public Policies and Private Troubles*, Cassell: London.

Ingold, T. (2011) "From trust to domination. An alternative history of human–animal relations," in Manning, A., Serpell, J. (eds.) *Animals and Human Society: Changing Perspective*, Routledge: New York: 1–22.

Instone, L., Mee, K. (2011) "Doggy encounters: performing new pet relations in the park," in Bull, J. (ed.) *Animal Movements. Moving Animals. Essays in Direction, Velocity and Agency in Humanimal Encounters*, Crossroads of Knowledge no. 17, Centre for Gender Research, Uppsala University: Uppsala: 229–250.

Irvine, L. (2003) "The problem of unwanted pets. A case study in how institutions 'think' about clients' needs," *Social Problems*, 50(4): 550–566.

Iversen, C. (2012) "Recordability: resistance and collusion in psychometric interviews with children," *Discourse Studies*, 14(6): 691–709.

Iversen, C. (2013) "'I don't know if I should believe him': knowledge and believability in interviews with children," *British Journal of Social Psychology*, 53(2): 367–386.

Jacobs, K., Gabriel, M. (2013) "Introduction: homes, objects and things," *Housing, Theory & Society,* 30(3): 213–218.

Jencks, C. (1994) *The Homeless*, Harvard University Press: Cambridge, MA.

Jerolmack, C. (2013) *The Global Pigeon*, Chicago University Press: Chicago, IL.

Jönsson, M. (2009) "Tantförvandlingen. Om kroppens längtan, skam och exotisering i Kerstin Thorvalls reseromaner," *Tidskrift för Genusvetenskap*, 4: 5–26.

Judd, D.R., Simpson, D. (2011) *City, Revisited. Urban Theory from Chicago, Los Angeles, and New York*, University of Minnesota Press: Minneapolis, MN.

Kaika, M. (2005) *City of Flows. Modernity, Nature, and the City*, Routledge: London.

Kennett, P., Marsh, A. (1999) *Homelessness: Exploring the New Terrain*, Policy Press: Bristol.

Kidder, J. (2011) *Urban Flow. Bike Messengers and the City*, Cornell University Press: Ithaca, NY.

Kittay, E.F. (1999) *Love's Labor. Essays on Women, Equality, and Dependency*, Routledge: New York.

Knutsson, G. (1939 [2014]) *Berättelser om Pelle Svanslös*, Bonnier Carlsen: Stockholm.

Krefting, E. (2009) "Animals in the city at the turn of the enlightenment: pictures from Louis Sébastien Mercier's Paris," in Holmberg, T. (ed.) *Investigating Human–Animal*

Relations—in Science, Culture and Work, Crossroads of Knowledge no. 10, Centre for Gender Research, Uppsala University: Uppsala: 33–43.

Kumlin, T. (2011) "Harold Garfinkel och den omedelbara sociala ordningen," in Lindblom, J., Stier, J. (eds.) *Det socialpsykologiska perspektivet*, Studentlitteratur: Lund.

Låt den rätte komma in (2008) (film), Svenska Filminstitutet: Stockholm.

Latour, B. (1988) *Science in Action*, Harvard University Press: Cambridge, MA.

Latour, B. (1993) *We Have Never been Modern*, Harvard University Press: Cambridge, MA.

Latour, B. (2005) *Reassembling the Social. An Introduction to Actor-Network-Theory*, Oxford University Press: Oxford.

Lefebvre, H. (1971) *Everyday Life in the Modern World*, Harper & Row: New York.

Lefebvre, H. (1991) *The Production of Space*, Blackwell Publishers: London.

Lefebvre, H. (1996) *Writings on Cities*, Blackwell Publishers Ltd: London.

Lefebvre, H. (2003 [1970]) *The Urban Revolution*, University of Minnesota Press: Minneapolis, MN.

Levine, D. (1972) *Georg Simmel on Individuality and Social Forms: Selected Writings by Georg Simmel*, University of Chicago Press: Chicago, IL.

Lighthouse Field Beach Rescue v. City of Santa Cruz, 131 Cal. App. 4th 1170 (2005)

Lindqvist, J.A. (2004) *Låt den rätte komma in*, Ordfront Förlag: Stockholm.

Lockwood, R. (1994) "The psychology of animal collectors," *American Animal Hospital Association Trends Magazine*, 9(6): 18–21.

Lönngren, A.S. (2015) *New Knowledges. Following the Animal in Literary Humanimal Transformations in Northern Europe After 1888*, Cambridge Scholars Publishing: Cambridge.

Lulka, D. (2010) "To turn: California's proposition 2 and the ethics of animal mobility in agriculture," *Humanimalia*, 2(1): 32–59.

Lykke, N. (2005) "Nya perspektiv på intersektionalitet. Problem och möjligheter," *Kvinnovetenskaplig Tidskrift*, 2–3: 7–17.

Lynch, M. (1993) *Scientific Practice and Ordinary Action: Ethnomethodology and Social Studies of Science*, Cambridge University Press: Cambridge.

Lyxfällan (2011) TV3 (TV series), retrieved from www.tv3play.se/program/lyxfallan/244516 (accessed December 18, 2014).

Markovits, A.S., Queen, R. (2009) "Women and the world of dog rescue. A case study of the state of Michigan," *Society and Animals*, 17(4): 325–342.

Mason, J. (2005) *Civilized Creatures. Urban Animals, Sentimental Culture, and American Literature, 1850–1900*, John Hopkins University Press: Baltimore, MD.

Mason, J., Tipper, B. (2008) "Being related: how children define and create kinship," *Childhood*, 15(4): 441–460.

Massey, D. (1994) *Space, Place and Gender*, University of Minnesota Press: Minneapolis, MN.

McAllister Groves, J. (1996) *Hearts and Minds. The Controversy over Laboratory Animals*, Temple University Press: Philadelphia, PA.

McCarthy, P. (2013) *Complex Pile* (art project).

McHugh, S. (2004) *Dog*, Reaktion Books: London.

McHugh, S. (2010) "Allowed," unpublished paper.

McHugh, S. (2011) *Animal Stories. Narrating Across Species Lines*, Minnesota University Press: Minneapolis, MN.

McHugh, S. (2012) "Bitch, bitch, bitch: Personal criticism, feminist theory, and dog-writing," *Hypatia*, 27(3): 616–635.

Michael, M. (2000) *Reconnecting Culture, Technology and Nature. From Society to Heterogeneity*, London: Routledge.

Miele, M. (forthcoming) "Cat communities," unpublished paper.

Miller, A. (2011) "Just don't call me 'mom': pros and cons of a family law model for companion animals in the US," *Humanimalia*, 2(2): 1–25.

Mouffe, C. (1999) "Deliberative democracy or agonistic pluralism," *Social Research*, 66(3): 745–758.

Mulkay, M. (1997) *The Embryo Research Debate*, Cambridge University Press: Cambridge.

Myers, O.E. (2003) "No longer the lonely species: a post-Mead perspective on animals and sociology," *International Journal of Sociology and Social Policy*, 23(3): 46–68.

Nathanson, J. (2009) "Animal hoarding: slipping into the darkness of comorbid animal and self neglect," *Journal of Elder Abuse & Neglect*, 21(4): 307–324.

Newton, M. (2002) *Savage Girls and Wild Boys*, Faber & Faber: London.

Oliver, K. (2009) *Animal Lessons. How They Teach us to be Human*, Columbia University Press: New York.

Österholm, M.M. (2012) *Ett flicklaboratorium i valda bitar. Skeva flickor i svenskspråkig prosa från 1980 till 2005*, Rosenlarv Förlag: Stockholm.

Oxford English Dictionary (2012) Online version, retrieved from www.oed.com (accessed December 18, 2014).

Park, R. (1915) "The city: suggestions for the investigation of human behavior in the city environment," *American Journal of Sociology*, 20(5): 577–612.

Park, R. (1928) "Human migration and the marginal man," *American Journal of Sociology*, 33(6): 881–893.

Park, R. (1967) *On Social Control and Collective Behavior*, University of Chicago Press: Chicago, IL.

Patronek, G. (1999) "Hoarding of animals: an under recognized public health problem in a difficult to study population," *Public Health Reports*, 11(4): 81–87.

Patronek, G. (2001) "The problem of animal hoarding," *Municipal Lawyer*, 19: 6–19.

Patronek, G. (2008) "Animal hoarding: a third dimension of animal abuse," in Ascione, F.R. (ed.) *International Handbook of Theory and Research on Animal Abuse and Cruelty*, Purdue University Press: West Lafayette, IN: 221–246.

Patronek, G.J., Glickman, L.T., Beck, A.M., McCabe, G.P., Ecker, C. (1996) "Risk factors for relinquishment of cats to an animal shelter," *Journal of the American Veterinary Medical Association*, 209(3): 582–588.

Patterson, M. (2002) "Walking the dog: an urban ethnography of owners and their dogs in the Glebe," *Alternate Routes*, 18: 5–70.

Peggs, K. (2012) *Animals and Sociology*, Palgrave Macmillan: Basingstoke.

Philo, C., Wilbert, C. (eds.) (2000) *Animal Spaces, Beastly Places. New Geographies of Human–Animal Relations*, Routledge: London.

Pink, S. (2009) *Doing Sensory Ethnography*, Sage: London.

Pink, S. (2011) "Multimodality, multisensoriality and ethnographic knowing: social semiotics and the phenomenology of perception," *Qualitative Research*, 11(3): 261–276.

Potter, J. *(*2010*)* "Contemporary discursive psychology: issues, prospects, and Corcoran's awkward ontology," *British Journal of Social Psychology*, 49: 657–678.

Rawls, A. (2002) "Editor's introduction," in Garfinkel, H. (ed.) *Ethnomethodology's Program: Working out Durkheim's Aphorism*, Rowman & Littlefield: Lanham.

Redmalm, D. (2013) "An animal without an animal within. The powers of pet keeping," PhD dissertation, Örebro University: Örebro.

Redmalm, D. (2014) "Holy bonsai wolves: Chihuahuas and the Paris Hilton syndrome," *International Journal of Cultural Studies*, 17(1): 93–109.

Reinisch, A.I. (2008) "Understanding the human aspects of animal hoarding," *Canadian Veterinary Journal*, 49(12): 1211–1214.

Rescue Santa Cruz Beaches (2004) (website), retrieved from www.thepetitionsite.com/takeaction (accessed June 11, 2010).

Robins, D., Sanders, C., Cahill, S. (1991) "Dogs and their people: pet-facilitated interaction in a public setting," *Journal of Contemporary Ethnography*, 20(3): 3–25.

Roe, E., Buller, H., Bull, J. (2011) "The performance of farm animal welfare," *Animal Welfare*, 20(1): 69–78.

Rogers, M.K. (2006) *Cat*, Reaktion Books: London.

Rudy, K. (2011) *Loving Animals*, Minnesota University Press: Minneapolis, MN.

Russell, N. (2007) "The domestication of anthropology," in Mullin, M., Cassidy, R. (eds.) *Where the Wild Things are Now*, Bloomsbury Academic: New York: 27–48.

Sabloff, A. (2001) *Reordering the Natural World. Humans and Animals in the City*, University of Toronto Press: Toronto.

Sahlin, I. (1992) "Hemlöshet: reflexioner kring ett begrepp," in Järvinen, M., Tigerstedt, C. (eds.) *Hemlöshet i Norden*, NAD-publikation 22: Nordiska nämnden för alkohol- och drogforskning: Helsinki: 95–122.

Sanders, C.R. (1999) *Understanding Dogs. Living and Working with Canine Companions*, Temple University Press: Philadelphia, PA.

Sanders, C.R. (2003) "Actions speak louder than words: close relationships between humans and nonhuman animals," *Symbolic Interaction*, 26(3): 405–426.

Segerdahl, P. (ed.) (2011) *Undisciplined Animals. Invitations to Animal Studies*, Cambridge Scholars Publishing: Cambridge.

Sennett, R. (1970) *The Uses of Disorder. Personal Identity and City Life,* W.W. Norton: New York.

Sennett, R. (1994) *Flesh and Stone. The Body and the City in Western Civilization*, Faber & Faber: London & Boston.

Serpell, J. (1986) *In the Company of Animals. A Study of Human–Animal Relationships*, Cambridge University Press: Cambridge.

Serres, M. (2008) *Five Senses. A Philosophy of Mingled Bodies*, Continuum International Publishing: London.

Shields, R. (1991) *Places on the Margin. Alternative Geographies of Modernity*, Routledge: New York.

Shields, R. (1998) *Lefebvre, Love and Struggle. Spatial Dialectics*, Routledge: London.

Simmel, G. (1950) *The Sociology of Georg Simmel*, The Free Press: New York.

Simmel, G. (1972) *On Individuality and Social Forms*, University of Chicago Press: Chicago, IL.

Simmel, G. (1981) *Hur är samhället möjligt?—och andra essäer*, Bokförlaget Korpen: Göteborg.

Simmel, G., Frisby, D.P., Featherstone, M. (1997) *Simmel on Culture. Selected Writings*, Sage: London.

Skeggs, B. (2000) *Att bli respektabel. Konstruktioner av klass och kön* [Formations of class and gender: becoming respectable], Daidalos: Göteborg.

Skeggs, B., Woods, H. (2011) *Reality Television and Class. International Perspectives*, BFI Publishing: London.

Smith, J.-A. (2003) "Beyond dominance and affection: living with rabbits in post-humanist households," *Society & Animals*, 11(2): 182–197.

Snæbjörnsdóttir, B., Wilson, M. (2009) "Radio animal," (art project), retrieved from www.snaebjornsdottirwilson.com/category/projects/radio-animal/ (accessed August 3, 2014).

Snæbjörnsdóttir, B., Wilson, M. (2011) *Uncertainty in the City*, Green Box: Berlin.

Somerville, P. (2013) "Understanding homelessness," *Housing, Theory & Society*, 30(4): 384–415.

South China Morning Post (2013) "Snip for strays helps win feral cat fight," 9 June.

Statistiska Centralbyrån (SCB) (2006) *Förekomst av sällskapsdjur—främst hund och katt—i svenska hushåll*, SCB Report: Stockholm.

Stokols, D. (1972) "On the distinction between density and crowding: some implications for future research," *Psychological Review*, 79(3): 275–278.

Stosuy, T. (2007) "At the Friends of the Lighthouse Field Community Forum 10 May," retrieved from www.youtube.com/watch?v=FQNoykXSxlk (accessed February 9, 2011).

Stratford, E. (2002) "On the edge: a tale of skaters and urban governance," *Social & Cultural Geography*, 3(2): 193–206.

Strindberg, A. (1889 [2007]) *Tschandala*, Nordiska Förlaget: Stockholm.

Superior Court of California, County of Santa Cruz (2005) Lighthouse Field beach rescue vs City of Santa Cruz et al., filed 10 August.

Swärd, H. (1998) *Hemlöshet. Fattigdomsbevis eller välfärdsdilemma?* Studentlitteratur: Lund.

Tarde, G. (2013) "Conviction and the crowd," *Distinktion. Scandinavian Journal of Social Theory*, 14(2): 232–239.

Thompson, K. (2007) *Performing Human–Animal Relations in Spain: An Anthropological Study of the Bullfighting from Horseback in Andalusia*. PhD dissertation, Discipline of Anthropology, University of Adelaide: Adelaide.

Tipper, R. (2010) "Moving encounters with pets: the secret life of Vince the cat," paper presented at the conference Animal Movements, Moving Animals, Centre for Gender Research, Uppsala University: Uppsala: 26–28 May.

Tissot, S. (2011) "Of dogs and men: the making of spatial boundaries in a gentrifying neighborhood," *City & Community*, 10(3): 265–284.

Tonkiss, F. (2005) *Space, the City and Social Theory. Social Relations and Urban Forms*, Polity Press: Cambridge.

Torgan, B.S. (2005) Re Lighthouse Field beach rescue v. City of Santa Cruz et al., State of California, *Dept. of Parks and Recreation*, 26 October.

Törnqvist, M. (2013) *Tourism and the Globalization of Emotions: The Intimate Economy of Tango*, Routledge: London.

Toth, J. (1993) *The Mole People. Life in the Tunnels Beneath New York City*, Chicago Review Press: Chicago, IL.

Tuan, Y.-F. (2005) *Dominance and Affection. The Making of Pets*, Yale University Press: New Haven, CN.

Uppsalatidningen (2011) "Här levde fler än 50 katter i misär" ["More than 50 cats lived here in misery"] (news article), 25 February.

Urbanik, J. (2009) "'Hooters for neuters': sexist or transgressive animal advocacy campaign?" *Humanimalia*, 1(1): 1–23.

Vaca-Guzman, M., Arluke, A. (2005) "Normalizing passive cruelty: the excuses and justifications of animal hoarders," *Anthrozoos*, 18(4): 338–357.

Valentine, G. (2004) *Public Space and the Culture of Childhood*, Ashgate: Aldershot.

van Doreen, T., Rose, D.B. (2012) "Storied-places in a multispecies city," *Humanimalia*, 3(2): 1–27.

Waddington, D.P. (2011) "Public order policing in South Yorkshire, 1984–2011: the case for a permissive approach to crowd control," *Contemporary Social Science*, 6(3): 309–324.

Walker, A. (2013) "'What can a crowd do?' Revisiting Tarde after the demise of the public," *Distinktion. Scandinavian Journal of Social Theory*, 14(2): 227–231.

Waltner-Toews, D. (2013) *The Origin of Feces*, ECW Press: Toronto.

Watson, S. (2006) *City Publics: The (Dis)Enchantments of Urban Encounters*, Routledge: London.

Weaver, H. (2013) "Becoming in kind: race, class, gender and nation in cultures of dog rescue and dog fighting," *American Quarterly Review*, 65(3): 689–709.

Weszkalnys, G. (2007) "The disintegration of a socialist exemplar: discourses on urban disorder in Alexanderplatz, Berlin," *Space and Culture*, 10(2): 207–230.

Wetherell, M., Potter, J. (1988) "Discourse analysis and the identification of interpretive repertoires," in C. Antaki (ed.) *Analysing Everyday Explanation: A Casebook of Methods*, Sage: Newbury Park, CA: 168–183.

Whatmore, S. (2002) *Hybrid Geographies. Natures, Cultures, Spaces*, Sage: London.

Whatmore, S. (2006) "Materialist returns. Practicing cultural geography in and for a more-than-human world," *Cultural Geographies*, 13(4): 600–609.

Wikipedia (2013) "Animal hoarding," retrieved from http://en.wikipedia.org/wiki/Animal_hoarding (accessed December 18, 2014).

Wiktionary (2014) "Underdog," retrieved from http://en.wiktionary.org/wiki/underdog#English (accessed December 18, 2014).

Wirth, L. (1938) "Urbanism as a way of life," *American Journal of Sociology*, 44(1): 1–24.

Wolch, J. (1998) "Zoöpolis," in Wolch, J., Emel, J. (eds.) *Animal Geographies: Place, Politics and Identity in the Nature–Culture Borderlands*, Verso: London.

Wolch, J. (2002) "Anima Urbis," *Progress in Human Geography*, 26(6): 721–742.

Wolch, J., Emel, J. (eds.) (1998) *Animal Geographies: Place, Politics and Identity in the Nature–Culture Borderlands*, Verso: London.

Wolch, J., Dear, M., Akita, A. (1988) "Explaining homelessness," *Journal of the American Planning Association*, 54(4): 443–453.

Wolch, J.R., West, K., Gaines, T.E. (1995) "Transspecies urban theory," *Environment and Planning D: Society and Space*, 13(6): 735–760.

Wolfe, C. (2003) *Animal Rites: American Culture, the Discourse of Species, and Posthumanist Theory*, University of Chicago Press: Chicago, IL.

Wolfe, C. (2009) *What is Posthumanism?* University of Minnesota Press: Minneapolis, MN.

Yelp (2006) *Lighthouse Field State Beach* (review by Tracey, W.), retrieved from www.yelp.com/biz/lighthouse-field-state-beach-santa-cruz (accessed August 18, 2010).

Yelp (2008) *Lighthouse Field State Beach* (review by Kristen, S.), retrieved from www.yelp.com/biz/lighthouse-field-state-beach-santa-cruz (accessed August 18, 2010).

Ytrehus, S. (2001) "Interpretation of housing needs? A critical discussion," *Housing Theory & Society*, 17(4): 166–174.

Zukin, S., Kasinitz, P., Chen, X. (eds.) (forthcoming) *Global Cities, Local Streets: Shopping in Place from New York to Shanghai*, Routledge: New York.

Index

ABC towns 107, 108
Acampora, Ralph 12
aesthetics 37
Africa 68, 122
Agamben, Giorgio 8, 133
age 100, 101
agency: cats 64, 110; hoarded animals 73; sentient 94, 96; social agency of non-humans 119; urban politics 126; zoöpolis 125
Aglert, Katja 19–20, 128–145
agonistic pluralism 19, 126–127, 142
Ahmed, Sarah 13, 28, 68
Alfredsson, Tomas 107
allowability 18, 23, 25, 36, 37, 43, 46
Amin, Ash 119, 122–123
Amsterdam 9
Andalusia 29
Andersson, Kay 3
animal hoarding 19, 71–96, 120–121, 130, 141; cat ladies 102, 104; definitions of 75, 79; detecting and recording 86–94; experience of trauma or abuse 77, 94–95, 110–111; explanations of 76–86; loss of control 84–86; love of animals 79–83, 94; as psychological disorder 76, 77, 111; rescue hoarders 78, 79, 82, 112–114; as a social problem 77–79; stigmatization 76, 81, 83, 90, 94; verminizing 95–96; victims and perpetrators 111–112
Animal Planet 72, 80–81, 85
animal rights 115, 126–127
animal studies 2–3, 5, 12, 15, 100, 144
Animal Welfare Act (1988) 48, 54–55, 63–64, 88

animal welfare officers 16, 74, 76, 87, 88–94
"animaling" 100, 114, 121, 122
animality 96, 110, 124, 133
"anthropo-zoo-genetic" practices 11, 59
anthropocentrism 5, 6, 8, 17, 18, 115
"anthropological machine" 8, 133
Aristocats (film) 47
Arluke, Arnold 77–78, 110–111, 112
art 128, 135, 136–137, 144
attachment model 77
Attenborough, David 141
Australia 32, 140

Bakhtin, Mikhail 44
Barad, Karen 143
Barry Island 33, 40
beaches 32, 45, 136
Becker, Howard 13
"becoming in kind" 104, 123
"becoming with" 8–9, 11, 52, 60, 105, 143
Berlin 4, 23, 137, 142–143
Bilbao 11
bio-politics 15, 39
birds 8, 30, 72–73, 138, 139
Birke, Lynda 14, 100, 114
Björklund, Anders 106–107
bodies 3, 5–6, 13, 120; crowding 124, 143; emplacement 95; interpretative repertoires 17; sociation 30
Borch, Christian 123
Botswana 122
boundaries 10, 27–28, 31, 136, 144
Bowen, James 59
Braidotti, Rosi 115
Brighenti, Andrea Mubi 15, 123

Brighton 32, 44
Broberg, Gunnar 47, 101
butchering 8
Butler, Judith 100

capitalism 124
carnivalesque 44, 45
"cat-egorization" 47, 50, 53–55, 66, 68,
    101, 114, 120, 138–140
Cat Ladies (film) 103–104, 105, 140
cats 2, 47–68, 101–102, 120, 138–140, 143;
    cat ladies 19, 97–116, 121, 132, 140; feral
    8, 9–10, 14, 18–19, 48–50, 52–64, 68,
    112–114, 120; hoarding 78, 79–81, 82–85,
    89, 91–93, 96, 111; homelessness 18, 48,
    51, 53–55, 65, 68, 120; liminal spaces 7,
    131; public/private distinction 136
Chicago School 10–11
chickens 122
cities 5, 10–11, 12, 119, 122, 125–126,
    136; *see also* urbanization; zoocities
civilization 138
Clark, T.N. 123
class: allowability 43; "becoming in kind"
    104; cat ladies 101, 105, 106, 114, 121;
    intersectionality 98, 100, 110; middle-class
    norms 96, 106; social construction of 114
co-evolution 2, 5, 11, 99, 141
collecting 72
collectives 14–18
companion species 1–2, 8, 97, 98, 99–100, 132
competition 5
complaints 88, 93
Complex Pile (art project) 136–137
Confessions (TV show) 72, 80–81, 85
conservation 37
contact zones 8–9, 87
control, loss of 84–86, 96, 114, 116
costs 75, 79
crowding 3, 5, 14–16, 19, 122, 142–143;
    animal hoarding 92, 96; cat ladies
    98, 116; cats 53, 65–66; numbers 40;
    promises of 123–125, 127
crowds 14–15, 44–45, 123–124, 142
cultural studies 12
cyborgs 131–132

deer 2, 6
Deleuze, Gilles 15

dependency 83
Despret, Vinciane 12, 59
deviance 13, 19, 76, 78–79, 81, 92, 94, 96,
    120
dialectics 6, 30, 32, 127
Dikeç, Mustafa 45
disciplinary power 58, 131
disorder 3, 20, 43, 45, 50, 127; humanimal
    crowding 98, 116; "insanity of place" 52;
    Western perspective 122; zoöpolis 125
displacement 68
diversity 126, 141–142
dog owners 36, 38
"dogginess" 39–41
dogs 1, 2, 10, 23–26, 30, 99, 120, 134–135,
    143; Bilbao 11; feces 136–137; hoarding
    71–72, 77–78, 81–82, 85–86, 88–90,
    111; interspecies identity 104; Kulik's
    work 134; liminal spaces 7, 31, 131;
    politics of place 43–45; regulation 6, 23,
    32–33; Santa Cruz dog beach 18, 24, 29,
    32–42, 43, 45; stray 57
domestication 2, 52, 96; cats 47, 58,
    59–60, 101; dogs 10, 23, 25
Donaldson, Sue 126–127
Donovan, Josephine 12
Douglas, Mary 7, 67, 132, 137
Drury, John 123
dualism 6

eccentricity 19, 106–107, 116
Elias, Norbert 138
emotions 12, 13, 28, 44, 134, 135; animal
    hoarding 95; cat ladies 115, 140;
    sociation 30
emplacement 52, 57, 74, 95, 121, 144
environmental protection 36–38
Ericksson, Barbara 77–78, 90
ethics 115
ethnomethodology 16–17, 74, 119, 131
euthanasia 57, 75, 82, 120
exceptionalism, human 5, 100, 116, 132, 141
exclusion 18, 19, 20, 25, 31, 43, 46, 120

Farsta 106–107
feces/excrement 20, 110, 136–137; animal
    hoarding 78, 81, 85, 86, 89, 96; Santa
    Cruz dog beach 34, 37, 38–39, 40,
    43–44

Feld, Steven 89
femininity 19, 66, 97, 101, 110, 114–116, 121, 141; *see also* women
feminism 100, 115, 116, 121, 143, 144
feral cats 8, 9–10, 14, 18–19, 48–50, 52–64, 68, 112–114, 120
"feral supporters" 66
figurative tropes 14
"fingeryeyes" 12, 87
Fiske, John 32
form 6, 26–27, 30, 32, 52, 107
Foucault, Michael 13, 24, 39, 131
Fox, Rebekah 50
foxes 14, 140
"frames of meaning" 2
Franklin, Adrian 10, 52
Franzén, Mats 29
Freud, Sigmund 65
Frost, R.O. 76, 77, 79, 86, 111
Fudge, Erica 10, 132

Gandy, Matthew 5, 39, 137
Garfinkel, Harold 16, 95, 129–130, 131
gender: allowability 43; animal hoarding 83; "becoming in kind" 104; cat ladies 97, 99–101, 102–103, 105, 106, 112, 114–115, 121; feline symbolism 47; intersectionality 98, 100, 110; social construction of 114; *see also* femininity; masculinity; women
Giddens, Anthony 10
Gieryn, Thomas 30
Goffman, Erving 2, 13, 76, 83
Gothenburg 29
governmentality 15, 24, 58
Griffiths, Huw 52, 66

Haraway, Donna 4, 12, 14, 39, 101, 130, 134, 144; animal rights discourses 115; "becoming with" 8–9, 60, 105, 143; cat's cradle 120; companion species 97, 99–100, 132; crowds 15; cyborgs 131–132; domestication 52; intersectionality 97, 98, 107; material-semiotic figurations 27; post-humanism 124; "response-ability" 75, 127, 141; vampires 109; zoöpolis 125
Hayward, Eva 8, 12, 87
health issues 39, 78

heteronormativity 116, 121
Hinchliffe, Steve 19, 125–126, 127
Hjerta, Elisabeth Margareta 101
Hoarders (TV show) 72, 86, 111
hoarding 19, 71–96, 120–121, 130, 141; cat ladies 102, 104; definitions of 75, 79; detecting and recording 86–94; experience of trauma or abuse 94–95, 110–111; loss of control 84–86; love of animals 79–83, 94; as psychological disorder 76, 77, 111; rescue hoarders 78, 79, 82, 112–114; as a social problem 77–79; stigmatization 76, 81, 83, 90, 94; verminizing 95–96; victims and perpetrators 111–112
Hoarding of Animals Research Consortium (HARC) 77
Holmberg, Tora 74
home 47, 50, 51, 73–74, 87, 96, 121, 136
"homeability" 60, 66, 67, 96, 114
homelessness 49–51, 59; animal hoarding 78; cats 18, 48, 51, 53–55, 65, 68, 120; definitions of 65; displacement 68
Hong Kong 62
Hovorka, Alice 122
Howell, Philip 24–25
Howes, David 87
human/animal relations 3, 5, 6, 7, 52, 119; hoarding 79, 83, 86, 95, 96; interpretative repertoires 17; norms of 116; paradoxes 75; "sensuous governance" 88; spatial conflicts 29–30
human exceptionalism 5, 100, 116, 132, 141
human-ness 5, 105, 110, 121, 141, 144
humanimal crowding *see* crowding
Humanimal group 17–18
humanism 7–8, 18
hunting 6
hybrid approach 5, 6

identity: animal hoarders 83, 95; cat ladies 97–98, 105; humanimal crowds 124; interspecies 104
inclusion 18, 23, 25, 31, 43, 46, 120, 130
indexicality 57, 92, 94, 129–130
Ingold, Tim 96
"insanity of place" 52
interpretative repertoires 17

intersectionality 43, 97, 98, 100, 107, 110, 121, 123, 141
interspecies identity 104
Its Beach, Santa Cruz 18, 24, 29, 32–42, 43, 45

Jencks, Christopher 50
Jerolmack, Colin 7, 30

Kidder, Jeffrey 31
Killeen, Celeste 77–78
Knudsen, B.T. 123
Knutsson, Gösta 101
Kulik, Oleg 134
Kymlicka, Will 126–127

Lanzarote 49
leashes 42
Lefebvre, Henri 12, 27, 31, 51–52, 121–122, 143
Let the Right One In (film) 107–110
Lighthouse Field Park 33, 34, 35, 41
liminal space 7, 27, 31–32, 37, 43, 45, 58, 131–132
Lindqvist, Johan Aijvide 107
livestock 122
London 29, 30, 31, 58–59
Lönngren, Ann-Sofie 71
"loose spaces" 30
lost cats 53–54, 68
love of animals 79–83, 94, 96, 111, 112, 115–116
Luxury Trap (TV show) 105–106
Lykke, Nina 100

Manila 67
masculinity 110, 115, 141
Massey, Doreen 50, 97
McCarthy, Paul 136–137
McHugh, Susan 23, 39
meaning-making 13, 17, 29–30, 46, 68
media 78, 99, 112
Michael, Mike 42
"more-than-human" 5, 6, 11, 46
Mouffe, Chantal 19, 45, 126, 142
movement 43–44
multi-species urbanization 1

Nathanson, Jane 77
nature 5, 140–141; nature/culture distinction 7, 8, 32, 38, 40, 96, 100

neglect 71, 74, 90, 93, 94, 96, 112
neighbors 74, 88, 93, 95, 138
neutering 53, 58, 61–63, 66, 68
New York City 58, 133
Newton, Michael 58
noble savage 58
normality 57
norms 29, 46, 66; animal hoarding 80, 81, 96, 141; cat ladies 19, 116, 121; gender 97; middle-class 96, 106; "sensuous governance" 88; stigmatization 76
numbers of animals 40, 90–92, 130

obsessive compulsive disorder (OCD) 77, 83
odors 89, 92–94, 95
order 20, 27, 43, 45, 129–130, 131; *see also* disorder
ownership 63–64, 65–66

parental attachment 77
Park, Robert 13
parks 10, 24
"passive cruelty" 83, 112
Patronek, Gary 76, 78, 79
performativity 100
pets 7, 10, 99–100, 131–132; animal hoarding 75; cat ladies 105; killing of unwanted 48; as members of the family 50
pigeons 30
places 13, 30, 45; cats 18, 68; crowds 124; place images 32; *see also* emplacement
play 44, 45
politics of conviviality 19, 125–126
politics of place 2, 3, 25–26, 29–31, 41, 43–46
pollution 8, 68, 93, 109, 136, 141
post-humanism 7–8, 100, 124, 141
power 52, 60, 142; disciplinary 58, 131; intersectionality 100, 110; knowledge and 76; love relationships 115; senses as source of 87–88; "sensuous governance" 74, 120–121
production 17
proximity 28, 52, 74, 99, 124, 125, 143
psychological disorders 76, 77, 111
public/private distinction 8, 126, 136
public space 25, 136–137

queer perspective 100, 104, 110, 116

rabbits 86
race: allowability 43; "becoming in kind" 104; intersectionality 97, 98
Radio Animal 135
rats 14, 138, 140
Rawls, Anne Warfield 17
recordability 91–92, 95
red foxes 140
Redmalm, David 7, 132
regulation 6; Animal Welfare Act 63–64, 88; cats 48, 65–66; dogs 23, 32–33
rescue hoarders 78, 79, 82, 112–114
resistance 44
"response-ability" 75, 127, 141
risk 35–36
Rose, Deborah Bird 1, 13, 45, 125
Rosenfeld, E. 76
Rudy, Kathy 83, 115
rules 13, 27, 32–33, 42, 43, 132, 136; *see also* regulation

safety 35–36
Sanders, Clinton 104
Santa Cruz dog beach 18, 24, 29, 32–42, 43, 45
Sennett, Richard 13, 43, 44
sense-making 17, 73, 74, 95, 120
senses 12, 74, 75, 87–88, 89, 92–94, 95–96, 120–121, 134
"sensuous epistemology of environments" 89, 90
"sensuous governance" 74, 88, 94–96, 120–121
sentimentality 115, 116, 141
Serres, Michel 87
service dogs 10
sex/gender distinction 100
sexual abuse 77, 110–111
sexuality 100, 104, 110, 121
sharing 45
shelters 51, 54, 56–57, 60, 65; animal hoarding 84, 85, 93; shelter workers 112–114
Shields, Rob 17, 31–32, 44, 45, 107
"significant otherness" 8
Simmel, Georg 10–11, 26–28, 31; sensory impressions 12, 87; sociation 28, 51, 98–99, 124; the stranger 13, 66
Singapore 9, 58–59
Singer, Peter 115

Smith, Julie-Anne 80, 130
Snæbjörnsdóttir, Bryndís 135
social control 42
social relations 28, 30, 97, 105
social spatiation 31–32, 107
sociality 13, 27, 99, 105, 116, 122
sociation 5, 11, 12, 28, 30, 98–99, 124
sociology 3, 6, 10–11, 28
Somerville, P. 51
space 26–28, 51–52, 119; crowds 124; social relations 97; social spatiation 31–32, 107; verminizing 139
spatialization 52
spaying 58, 62; *see also* neutering
species 8, 15, 130; *see also* companion species
Steel, Carolyn 137
Steketee, G. 76, 79, 86, 111
stigmatization 19, 76, 81, 83, 90, 94
Stockholm 29, 72, 106–109, 130, 132, 138
storying 13, 45, 125
Stott, C. 123
strangers 13–14, 18–19, 50, 66, 68, 95, 120
Strindberg, August 71, 73
subjectivity 5, 6, 11, 52, 144; animal hoarding 72; crowds 124; dialectics 30, 32; social spatiation 107
suffering 75, 78, 79
Suomenlinna 129
surveillance 15
swans 72–73
Sweden: animal hoarding 72–73, 76, 78, 112, 130; animal welfare officers 16; cat ladies 99, 105–109; cats 18, 48, 51, 52–57, 62, 63–64, 65–66, 101; complaints about animals 88; dogs 1, 23; red foxes 140; twins 68; whiteness 143
Swift, Jonathan 58
Sydney 29
symbolism 10, 47, 58, 101

taming 58, 59–61
Tarde, G. 123
taxonomies 129
Thailand 26
third space 132–133
Tissot, Sylvie 24
trans-species urban crowds 44–45
trap-neuter-release (TNR) method 61–63, 68

traumatic crisis 94–95, 110–111
twins 68

"underdogs" 109
"unheimlich" (the uncanny) 64–65
United Kingdom 65
United States 48, 51, 65, 78; *see also*
    Santa Cruz dog beach
Uppsala 2, 17, 101
urban planning 15–16, 107
urban space 12–13, 26–27; *see also* cities;
    space
urban theory 5, 6–7
Urbanik, Julie 48
urbanism 12
urbanization 1–2, 5, 7, 10, 15, 23, 122

vampires 109–110
van Doreen, Thomas 1, 13, 45, 125
Venice 29, 30
verminization 16, 19, 74, 86, 95–96, 121,
    124, 138–141
victims, cat ladies as 110–112, 114
Vienna 23
Villesen, Gitte 144
virtual objects 14, 122

Waddington, D.P. 123
Walsh, Katie 50
Waltner-Toews, David 137
Ward, Kate 71–72, 73, 75
waste management 138
Watson, Sophie 136
Weaver, Harlan 104, 123
Whatmore, Sarah 19, 125–126,
    127
wild animals 1
wildlife 36–37
Wilson, Mark 135
Wirth, Louis 12
Wolch, Jennifer 6, 19, 125, 126
women: animal hoarders 77, 78,
    79–83, 111; cat ladies 19, 97–116,
    121, 132, 140; *see also* femininity;
    gender

zoocentrism 64
zoocities 3, 122, 126, 127, 128–129,
    130
zooethnography 16–18
zoöpolis 125, 126
zoos 3, 27
Zukin, Sharon 27